The Challenges of Revitalizing Rulal Areas in Japan

The Challenges of Revitalizing Rural Areas in Japan

Case studies by trailblazers

Supervised by The Rural Revitalization Trailblazers Conference
Edited by EY Strategy and Consulting

the japan times PUBLISHING

The Challenges of Revitalizing Rural Areas in Japan
Case studies by trailblazers

2025年5月10日　初版第1刷発行

監　修：地方創生先駆者会議
編　者：EYストラテジー・アンド・コンサルティング
発行者：伊藤秀樹
発行所：株式会社ジャパンタイムズ出版
〒102-0082 東京都千代田区一番町2-2 一番町第二TGビル2F
ISBN 978-4-7890-1922-4

Originally published in Japanese as *Chiho Sosei Senkusha Moderu* by Chuo Keizaisha in 2023.

First edition: May 2025

Supervised by The Rural Revitalization Trailblazers Conference
Edited by EY Strategy and Consulting
Editing and copyreading: Time & Space, Inc.
Cover art: Tetsuya Hagiwara
Editorial design and typesetting: Soju Ltd.
Published by The Japan Times Publishing, Ltd.
2F Ichibancho Daini TG Bldg., 2-2 Ichibancho, Chiyoda-ku, Tokyo 102-0082, Japan

Website: https://jtpublishing.co.jp/

Printed in Japan
ISBN 978-4-7890-1922-4

Table of Contents

This map shows towns and cities featured in this book.

Introduction: Origins of the Rural Revitalization Trailblazers Conference

Is rural revitalization dependent on a specific person?

After the Rural Revitalization Trailblazers Conference Symposium hosted by the EY Intelligence Platform on January 20, 2023, a participant offered the following feedback:

> I am in charge of rural revitalization at the local government level. I have been pretty much left to my own devices. But no one seems to understand what I am doing, and I have failed to produce any significant results. This has led me to wonder how my situation differs from regions that are successful. I am not sure how to succeed or where to find the answers I am looking for.
>
> The successful regions seem to be involving people who can be referred to as "super-specialists in rural revitalization." I had almost given up, because I thought that unless I could meet people like that, I would not make any progress. At least, that was how I felt until I came here today. Now I feel a lot better. I've started to feel like I can make a go of it.

There must be many people all over Japan who are in a similar position and are facing similar obstacles. The Rural Revitalization Trailblazers Conference was set up about two years ago in the hope of connecting such people and sharing know-how and insight to ensure the success of rural revitalization projects and, by extension, invigorate Japan again from the regional level.

The Japanese term "*chiho sosei,*" meaning rural revitalization in the sense of overcoming population decline and stimulating the local economy, was first used in September 2014. A rural revitalization policy was announced on the inauguration of the second Shinzo Abe cabinet, and Shigeru Ishiba was appointed as minister in charge of overcoming population decline and vitalizing the local economy, along with the establishment of the Headquarters for Overcoming Population Decline and Vitalizing the Local Economy. The purpose of this policy is to effectively address the challenges Japan faces in terms of its declining birthrate and aging population, halt population decline, and create an autonomous and sustainable society in which each region makes the most of its unique characteristics while balancing the concentration of the population in the Tokyo metropolitan area.

In the nine or so years since the policy was introduced, many regions have continued to implement various initiatives, using grants for promoting rural revitalization and the Hometown Tax Program. Nonetheless, there are still few examples that can really be called success stories.

There may be any number of reasons for this, but in most cases, it is due to a lack of human capital and expertise. Qualified personnel may be readily available in big companies situated in urban areas, but they are hard to come by in more regional areas. This means that when grants expire and a project comes to an end, the skilled personnel involved also return to the big companies, leaving nothing behind for the local community. No matter how many times a grant project is repeated, the outcome is always the same.

Rural areas lack the know-how and funds to take over such projects and develop them into sustainable businesses. It is difficult to achieve autonomous and sustainable local communities by simply continuing one-off projects that rely on grants.

On the other hand, numerous successful examples do exist, such as the initiatives by the trailblazers introduced in this book. What started out as a small project begun by a small group of private-sector volunteers soon drew attention and attracted both people from the local community and from outside the region, resulting in the launch of a number of projects. The main players are locals or people from outside the region who have moved to the area to become locals. These projects will continue even after help from big companies dries up. Possibly, the big companies will maintain their interest and commitment to the projects. There are various patterns to these success stories. In some cases, private-sector projects were the starting point, while in others, local government projects led to the development of other projects. Either way, independent efforts are spreading in these communities like wildfire.

The question is, where does this difference come from? And that is the theme of this book.

Most of these success stories have a central figure: a virtuoso who appears to be the "super-specialist in rural revitalization" that the symposium participant quoted above was referring to, or a group acting as the driving force, seemingly drumming up excitement from the outset. From afar, it would appear that the success of the region can be attributed to the actions of such people.

This has led to the theory that successful rural revitalization is dependent on a specific person. Charismatic individuals who want to revitalize the region, and who have the insight, vision, and energy to paint a specific picture, the knowledge and connections to raise money, the ability to start a business, and the ability to get others involved, are few and far between.

If revitalization were limited only to those regions fortunate enough to encounter such an individual, rural revitalization would be no more than a pipe dream.

Such a situation is far from ideal. Since the future of Japan is at stake, we cannot let rural revitalization be a mere coincidence. We cannot allow ourselves to fall into a situation in which regions feel fortunate that things look as if they will work out because they just happened to encounter an amazing individual.

So what should be done? We reached the following conclusion.

Looking closely at some of the success stories, which at a glance seem to be dependent on the expertise of a specific person or people, a certain model starts to emerge. "Amazing individuals" have unique personalities and skills which can be broken down into component elements. This in turn enables us to structure and formulate the success factors. By systematizing this as an empirical theory and developing it horizontally, we may be able to artificially reproduce a model of successful rural revitalization. The question is whether a rural revitalization start-up ecosystem can be artificially created.

To answer this question, it is essential to work with professionals who have a track record of successfully implementing numerous rural revitalization projects. These are people who have concrete know-how and knowledge based on their own successes, who are also capable of analyzing success stories in other regions and verbalizing them into abstract logic.

We refer to such people, who can move freely between the material and abstract worlds, as "trailblazers," and we have decided to bring them together for further discussions. We need to start by identifying a model.

That brings us to the origins of the Rural Revitalization Trailblazers Conference.

Building a new social system based on mutual aid

Our story began nearly three years ago. We, the members of the Government and Infrastructure Sector at EY Strategy and Consulting (EYSC), had been discussing how rural revitalization should take place, as described above.

EY, a professional firm with more than 700 offices in over 150 countries and regions worldwide, provides business consulting and other services to corporations and other organizations. So why is EY getting involved in rural revitalization? The answer is simple. We have established "Building a better working world" as a common purpose for all our member firms worldwide, and we place the highest priority on pursuing "long-term value."

The primary mission of consulting firms is usually to solve the business challenges facing their clients. At EY, this is no different, but in our case, we believe that beyond such management issues are problems facing each industry as a whole, and beyond that, there are social issues facing Japan. We recognize that our mission, in accordance with our purpose, is to help solve all of these issues.

Therefore, at EY we are not concerned only with immediate profits; rather, we are committed to taking a long-term perspective. We believe that rural revitalization is a topic that should be faced head-on by those of us who have an awareness of the issues involved, a sense of values, and a long-term perspective.

For us, rural revitalization is not the only objective. Our ultimate goal is to build a new social system for the whole of Japan, starting with regional transformation. In Japan, birthrates are declining, the population is aging, climate change issues are becoming more serious, and the threat of pandemics and geopolitical problems are increasing. At the same time, new social issues are emerging, including creating support systems for the elderly in regional communities, creating an environment where single-parent households can work and live without experiencing hardship, and providing small and medium-sized businesses with support for digital transformation.

Regional economies are the most seriously affected by these universal burdens in Japan. Increasing pressure is being placed on local government finances. Private-sector services are also being hampered by geographical diversity and dispersion, as well as low service-unit cost, and the sustainability of these services is being questioned. If issues that both government and private-sector services find difficult to solve continue to worsen, it will be up to the regions to identify a new mechanism to integrate the knowledge of both the government and the private sector and provide services to local residents through an organic division of roles between these sectors.

In other words, in addition to public aid from the government and independent financing from the private sector, a system of mutual aid through collaboration between the two is required. We believe that the establishment of such a system is the most urgent issue facing Japan.

Based on this vision, we have created a framework for the public and private sectors to work together to solve social issues and to create a society sustained by mutual aid by bringing together the insight of both parties. That framework is the EY Intelligence Platform, which we launched on November 12, 2021. We have positioned rural revitalization as one of the platform's key themes.

The EY Intelligence Platform consists of three subcommittees: the Trailblazers Conference, Thought Leadership, and Field Work.

Aiming to "spin out a core strategy," the Trailblazers Conference will verbalize a methodology for solving social issues, using advanced knowledge and skills that draw on both concrete and abstract strategies. Thought Leadership is a specialized subcommittee tasked with formulating the core know-how based on the central strategy. Thought Leadership will organize the components of the social system derived by the Trailblazers Conference and systematize them by adding new research and analysis. Then, we will move on to the Field Work team, which will take on the challenge of social implementation of the systematized mechanism through public-private partnerships at each worksite while eliciting cooperation from local governments in specific fields.

Origins of the Rural Revitalization Trailblazers Conference

The initial theme of the Trailblazers Conference was rural revitalization. The conference held its first session on November 12, 2021, the same day on which the EY Intelligence Platform was launched. The following is a list of conference members.

Trailblazers:
Toshiki Abe (representative director, Ridilover Inc.)
Kumi Fujisawa (chairperson, Institute for International Socio-Economic Studies, Ltd.)
Hima Furuta (representative director, Umari Inc.)
Shintaro Kado (president & representative director, Machidukuri-Matsuyama Co., Ltd.)
Yasuhiro Kamiyama (president and representative director, Hyakusenrenma, Inc.)
Daisuke Maki (president & representative director, A-Zero Group Inc.)
Keisuke Murakami (director-general, Digital Agency, Government of Japan)
Yoshiteru Takemoto (representative director, tobimushi Inc.)
Facilitators:
Jun Hori (representative director, Watashi o Kotoba ni Suru Research Institute)
Mayuko Miyase (director, Watashi o Kotoba ni Suru Research Institute)

The eight trailblazers we invited are all rural revitalization experts with outstanding track records of success in regional revitalization. Moreover, as mentioned above, we selected individuals who are not only able to talk about concrete examples and know-how based on their own experiences, but are also skilled in grasping the elements that can and cannot be seen and deriving abstract theories from such elements.

We asked Keisuke Murakami to lead the discussion among the trailblazers, and Jun Hori (former NHK announcer) and Mayuko Miyase (former Fuji Television announcer), both from Watashi o Kotoba ni Suru Research Institute, facilitated the sessions. Mr. Hori and Ms. Miyase produced informational videos on the case studies; these were

A scene from a session of the Rural Revitalization Trailblazers Conference.

The Conference members pose for a group photo. Left to right: Keisuke Murakami, Daisuke Maki, Toshiki Abe, Shintaro Kado, Hima Furuta, Kumi Fujisawa, Yasuhiro Kamiyama, Yoshiteru Takemoto, Mayuko Miyase, and Jun Hori.

shown at this conference and contributed greatly to the discussions.

A total of seven sessions were held in which seven of the trailblazers took turns giving presentations, starting with Mr. Hima Furuta at the first session and ending with Mr. Daisuke Maki at the last session on December 12, 2022.

Sessions were held approximately once every two months; each followed the same basic structure, starting with a presentation by a trailblazer to introduce an example of a rural revitalization project in which he or she had been involved. Video footage of actual on-site visits and interviews by Mr. Hori and Ms. Miyase were shown during breaks in the presentations. Following the presentations, the group discussed the elements that had contributed to the success of the projects, the challenges that had been overcome, any aspects that could be universalized, and what action would be taken in the future.

Prior to each session, Mr. Murakami, Mr. Hori, and Ms. Miyase conducted in-depth interviews with the presenting trailblazers to confirm the basic focus of their presentations, the key elements, and hints for their on-location interviews.

For example, the underlying theme of session 1, in which Mr. Furuta introduced a case study involving Mitoyo in Kagawa Prefecture, was "development from independent financing to mutual aid," starting with small projects at the private-sector level. In response, Yoshiteru Takemoto, the presenter at session 2, introduced "the flow from public aid to mutual aid," starting with the revitalization of a public high school that was about to close in the town of Ama in Shimane Prefecture. Session 3 took a completely different direction. Yasuhiro Kamiyama, a multitalented individual whose lobbying efforts led to the introduction of the Private Lodging Business Act, captivated us with his presentation on how to move politicians and bureaucrats.

Although the main theme of each session had been established in advance, various keywords were tossed around during the actual sessions and, as if inspired by these keywords, concepts that had not previously been explicitly stated were newly formulated in a common language.

For example, when activities in a certain area reach a certain stage, they will catch the attention of people from outside the area who will visit to see what they can do to help. In order to invite such people, who are initially mere spectators, to get involved in a specific activity, you need to encourage them to take your activities seriously, and they need roles to play and places to belong for their involvement to materialize and be long-lasting. However, this often leads to some kind of confrontation, and so a safety net should be put in place to protect newcomers. It is important to create a "kelp forest" system (see chapter 1, session 6 for details) for them to grow and stand on their own feet while gaining experience.

As key businesses that involve confluence, conflict, and growth emerge in various areas, they form a multilayered structure like a *mille-feuille* pastry before finally reaching the stage of full-scale rural revitalization.

The first chapter of this book recounts each of these sessions virtually in their entirety. This represents the development process of the discussions; at the same time, it outlines the process by which a model of rural revitalization, which emerges only when looking at the entire process from the first to the last session, gradually takes shape. We encourage you to read in order from the first session.

The complete minutes of each session are also available on the *Ridilover Journal*, a media site operated by trailblazer Toshiki Abe's company. Please refer to these minutes if you would like to find out more about the exchanges that took place during the sessions.

Evolving the Trailblazers Conference, aiming to promote rural revitalization

What we know now, after the seven sessions of Trailblazers Conference, is that we must keep on going.

Thanks to the discussions, we have been successful to a certain extent in identifying and articulating the elements that will lead to successful rural revitalization, and in creating a "model" of a theoretical system based on such elements. This is summarized by Mr. Murakami in chapter 2 as the "Rural Revitalization Trailblazing Model."

However, we have also identified the next set of issues that need to be further refined. One of these is human resource development. For this conference, we identified a need for "General Partners" (GP), borrowing a term from the finance world. GP, who act as the driving force, taking unlimited liability to run local businesses, are essential for rural revitalization. "Area organizers," who offer support from the local side, and "accelerators," who support the creation of a business environment—including financing—from the outside, are also necessary. Truth be told, in regional areas there is an overwhelming lack of people who are capable of taking on and supporting such commercialization.

In order to overcome this, the Trailblazers Conference has decided to establish a council of about 50 trailblazers and potential trailblazers to develop the people who will eventually become GP candidates, area organizers, and accelerators. Efforts have already begun, and if we consider the trailblazers involved in this Trailblazers Conference as the first generation, then we have just started to attempt to create the second and third generations. We hope to get this activity off the ground and set a new stage for rural revitalization using our networks of contacts.

"Is rural revitalization just dependent on a specific person?" The answer has to be no. That being said, we have come to realize through the Trailblazers Conference that the main support for rural revitalization comes from people, and that sustainability cannot be achieved without a posse of like-minded individuals who support each other.

As Shintaro Kado, who shared his struggles in his hometown of Matsuyama during session 5, noted, "At the end of the day, it's the people." Mr. Takemoto made the following comment during an interview for this book.

> [Even if we can't copy exactly what Mr. Furuta is doing,] by breaking down the functions of Mr. Furuta into their component elements. And if we look at Mr. Kado, the responsibilities that he has had to carry alone among several people. . . . However, that doesn't mean that we are going to dismiss the idea of relying on people. What I'm saying is that even though we don't necessarily need a Mr. X, we still need other people who can perform the functions that Mr. X served. In that sense, it might be better for us to say that when it comes to rural revitalization, we should not be looking at proper names (Mr. X), but rather specific people who can *function* as Mr. X.

As you will discover when you read chapter 3, the trailblazers are unanimous in testifying that the video footage taken by Mr. Hori and Ms.

Miyase was extremely meaningful in providing clarity to previously vague perspectives and ideas, and in creating a common language. The footage also seemed to be effective in helping everyone understand the situation in each region with their hearts rather than their heads. Some of the footage can be viewed in digest form on a special page on the *Ridilover Journal* website (see note).

This goes to show that emotional devices that appeal to people's hearts and minds are also essential for rural revitalization. Mr. Abe alludes to this as the need for a "story." In this sense, people will still play a leading role, and human resource development will have to wait. Something that lights a fire in the heart of a person—like the symposium participant who told us that he still feels like he can be successful—and then, like a bonfire, grows bigger as it connects to various people, will be the driving force that moves the entire system.

One more point. There are still some challenges for the next stage of this conference, which Mr. Maki talked about during the final session. Since choosing not to merge with neighboring municipalities 20 years ago, the village of Nishiawakura in Okayama Prefecture, where Mr. Maki operates, has focused on developing local ventures centered around reforestation projects, and has successfully spawned more than 50 start-ups. However, in order for each start-up to become a solid and sustainable business, it needs to grow above a certain size. Scaling up these businesses is the next challenge. Yet, if capital from outside the region is relied on to scale up operations, the businesses that locals have worked so hard to build up will end up without a local presence. That said, even if they wanted to go it alone, the region just does not have that kind of capital. This contradiction has caused many "successful" examples of rural revitalization initiatives to come to a standstill.

In session 4 of the Trailblazers Conference, Kumi Fujisawa pointed out the importance of shifting from a distribution mindset to an investment mindset, and noted that finance mechanisms are the key to expanding scale. This leads us to the question of what specific action should to be taken. In order to tackle this question, EY has decided to embark on a second round of the Trailblazers Conference during this fiscal year.

The first round could not identify all of the steps that will lead to successful rural revitalization. While it is undeniable that many questions and issues remain, we are convinced that one of the paths to the "summit" is beginning to clear, as if a fog were lifting.

In this book, we have tried to convey what really took place at each session, including the nuance and atmosphere, so that we can share as much as possible about the process that led us to this point. We have tried to use as many phrases and keywords as possible to create a common language that will help those who are actually involved in rural revitalization and those who support such revitalization from outside the region.

We hope that this book and the activities of the EY Intelligence Platform will contribute to the formation of a network to support rural revitalization going forward, as well as the creation of a rural revitalization start-up ecosystem.

EY Intelligence Platform Executive Office
September 2023

Chapter 1

Trailblazers Conference

Session 1: Mr. Hima Furuta's Case —Mitoyo, Kagawa Prefecture

From Udon House to Urashima Village: The number of tourists increased a hundredfold in five years

Mitoyo, a small seaside town in Kagawa Prefecture with a population of over 60,000 people, has come into the limelight as a successful case of rural revitalization. The cascade of new initiatives that began at Udon House gave rise to the buzz around Chichibugahama Beach—also known as Japan's "Salar de Uyuni" (a reference to the famed Bolivian salt flats)—on social media, attracting people and spawning various start-ups, with over 15 new companies already in motion. The first session of the Trailblazers Conference kicked off with a presentation by Hima Furuta of umari Inc., who played a part in driving the production of local businesses in Mitoyo.

Production of local businesses

Furuta: When we talk about rural revitalization, most cases often start by addressing local issues. But for me, I've always gone with producing businesses in the local community based on the idea that I want to build a town like this with everyone. And Mitoyo in Kagawa Prefecture is one such case.

When it comes to producing local businesses, the key is to prioritize community. People often say "global versus local," but if we view a region through this dichotomy, we will not see the same activities everywhere in the country. Global values are generally based on the logic of numbers. However, the viability or the strength of a local area does not lie in numbers. It is the community that drives that. With this in mind, we came to the decision that the dichotomy is not between global and local, but global and community. Creating attractive communities is the basis for producing local businesses.

Not high value-adds, but other value-adds

When you think about it this way, you are able to then see things that have the potential to be overwhelmingly popular with a certain group of customers, even if they are of no value to the majority of people. A run-down, weathered shack on the beach, while completely unattractive in terms of real-estate value from a global angle, may very well be valuable to surfers. We then thought that some kind of specialization could be used as a breakthrough to create a community that would cater to the different aspirations of each individual.

Another way we can look at it is additional value-adds rather than high value-adds. For example, many people will opt to provide more value-add products or services, such as selling juice made from the finest apples grown without pesticides for 10,000 yen a bottle. And indeed, this is likely to attract interest. But would it last long? Probably not. So, what exactly is an additional value-add? In Mitoyo, we have succeeded in turning udon (Japanese wheat noodles) into an educational kit that teaches people how to make noodles and more about udon, rather than just a dish to eat. In this way, we have added another value to udon noodles that is different from that of food.

To begin with, Kagawa Prefecture has already done their best in terms of product development and branding for udon noodles. After all these years, we did not think that pushing a new, high value-added udon priced at 1,000 yen would sell well. We therefore came up with the "Sanuki udon noodle educational kit for gifted children." It was a set that taught children how to make udon noodles right from the first step of using flour; we priced it at 7,000 yen for ten balls of noodles. This set was very popular with elderly folks who wanted to make udon noodles with their grandchildren or give it to them as a gift, and it sold like hotcakes. The udon-noodle set came up to 700 yen for each ball of noodles, which is indeed expensive for udon noodles. But what the customers bought from us was not just udon noodles; what we have put into their hands is time with their grandchildren, and also education, which are other values, different from the value of udon as food, that we have added.

Udon House and Chichibugahama Beach

I'll talk more about the initiatives that we implemented in Mitoyo one by one, in order.

We started with "Udon House." We bought a *kominka* (old traditional

Udon House is an experimental lodging facility where foreigners can stay.

Japanese house) and created an experiential lodging facility for foreigners to learn about and make Sanuki udon noodles. The concept is "Bed and Udon." Just like there are mountain huts where you can stay overnight in the mountains, we thought it would be interesting to have an udon shop where people can stay overnight as guests in the home of udon noodles. Udon House is a lodging business born from this idea, where visitors can learn about the local food culture.

A two-day, one-night stay is 30,000 yen. It may sound expensive, but what if, for the same price, you could, say, stay at a pizzeria in Naples, Italy? And not just stay there, but also learn how to make real pizzas and even interact with the locals. How does that sound? If anything, doesn't it feel inexpensive?

As a matter of fact, this Udon House caught the attention of foreign media and went viral; it was even featured on CNN Travel for ten days. Soon after, people from around 25 countries and regions began paying visits.

The next step was to create a place for more people to get involved and play an active role. Chichibugahama Beach is now a popular tourist spot known as Japan's "Salar de Uyuni," a reference to the famous Bolivian salt flat, but back then, it was really just a desolate beach that few people visited. But thanks to the locals who have been cleaning the

Chichibugahama Beach is a popular tourist spot known as Japan's "Salar de Uyuni" salt flats.

beach every month for over three decades, it is now a beach where one can walk barefoot. And it was around the time when Instagram started to become popular that people started posting about Chichibugahama Beach on social media.

When we heard that the designated administrator of the run-down public restrooms at Chichibugahama Beach was to be replaced, we proposed to implement it as Park-PFI project to the government. "We won't charge any designated administration fees. As we pay annual land rent, allow us do business here all year round." We proposed such a model with local companies and non-local companies specializing in maintaining green spaces, as well as local trading companies.

Based on this business model, we conducted workshops; the third-generation owner of a local supermarket opened a café to help the local community to connect within and beyond the region. This became the talk of the town, and more and more people started coming forward in their desire to want to contribute and do something. And with the explosion of Instagram, Chichibugahama Beach quickly became popular across Japan. This trend was not limited to Chichibugahama Beach; other players in the local community also started to form teams and created their projects. There's "Omusubi-za"—a tatami-floored play café, revamped from a *kominka*, started by a team in their 30s who were busy raising children. There's also a sake brewery, defunct for nearly 30 years, that was revived and transformed into a company-run restaurant-cum-accommodations by the fourth- and fifth-generation owners of a local century-old company. In this way, over 15 companies were launched one after another, with 30 projects eventually set in motion, transforming all of Mitoyo into a bustling region.

As the activities expanded and flourished in this way, more and more people from outside the prefecture came to us and asked to be a part of it as well. Since there were not enough places for these people to stay, we created a lodging facility called "Gate." This place does not provide just lodging, but also offers work and community. Gate introduces its guests to opportunities to find local work and interact with the community, such as helping out at a café on weekends or providing IT-related support, making it possible for them to stay for a long period of time. It is a system in which newfound local friends can support those who feel helpless and anxious about starting out in a new place for the first time.

In short, I believe that what we are creating in the community is not content, but function.

Dependency is the biggest enemy of the region

Let's move on to finances now. Before we start any project, we usually try to raise funds by investing in it ourselves or through crowdfunding. We can also go down the grants route, but things usually don't pan out if we depend on them.

I personally think that dependency is the biggest enemy of the region. It is important to see rural revitalization as something we need to be responsible for, and to create projects while supporting each other on our own. Instead of depending on the government or leaving it to large corporations or the bigwigs in town, we should adopt a self-reliant attitude—we can just create whatever is not available in the community. And crowdfunding is a good way to raise funds for such ideas.

In fact, once the number of tourists reaches about 500,000, large corporations begin to take notice. There have been rumors of large corporations planning to invest in new hotels, and some have the idea that the ensuing jobs created, if any, would be welcome news. But this would mean that the local companies would end up getting subcontracted work from the large corporations instead. When this came up as a topic over drinks one night with my Udon House partners, we decided that if that was the case, we might as well start our own hotel instead.

By this time, the scale of our projects had grown to the extent that we could see some success. So we decided to fund the project ourselves and take out a loan instead of relying on crowdfunding. To give you some precise numbers, 11 local member companies put in 5 million yen each, and we borrowed another 140 million yen from local financial institutions. The 11 companies included construction companies, building materials suppliers, furniture suppliers, bus companies, and power companies. These companies that invested in the hotel would be responsible for all the work and operations that would go into building and running the hotel. I believe that we have come up with a new model for regional development, in which the profits earned by the community are accrued and circulated within the community.

In recognition of the system based on our model, the Urashima Village hotel, which we built in a location famous for the legend of Urashima Taro, received the Best Award (Minister of Agriculture, Forestry and Fisheries Award)—Japan Wood Design Award—in 2021.

What matters is not the idea itself, but the reason

Before I conclude, I'd like to talk about two things that I have once again found to be important in the course of driving these projects.

The first thing is how to go about increasing the number of people involved or connected minds, and how to be on closer terms with them. The city of Mitoyo was formed by a merger of seven municipalities, and this meant that the identity of "Mitoyo" was somewhat lacking. In fact, people did not really know each other. And I think this was why we were able to inspire each other when talking about our vision, and even accepted "outsiders" from Tokyo without any prejudice. All these factors helped to fuel our feelings for our hometown.

It is essential to create a place where we can seriously discuss what we want to be or achieve. While this is not something that is formal or visible, it will inevitably provide significant support for the activities.

The other thing is to clearly separate "concepts" from "ideas." Concepts are "Why are we doing it?" and ideas are "How are we going to do it?"

Fig. 1-1. Factors required for rural revitalization

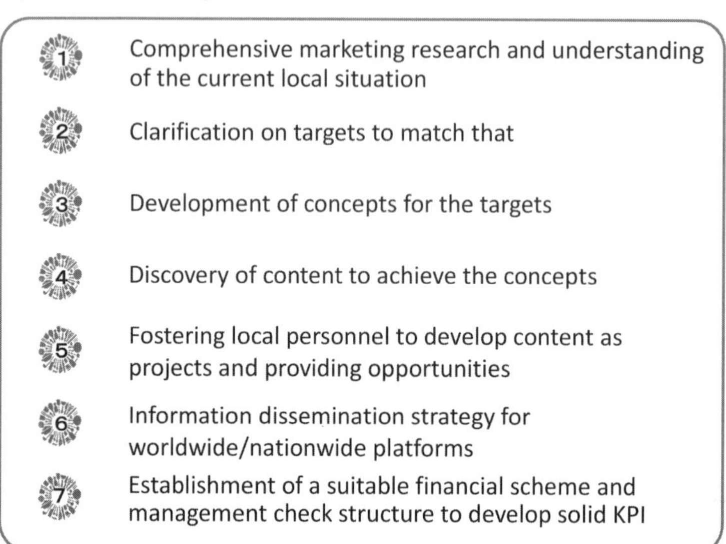

Rather than "just some" of the factors above, it is meaningless unless all of the factors together are supported as knowledge.

But we often see cases where the concept and idea are viewed as the same thing, or only one or the other is there.

Some examples that come to my mind now are local gourmet food, mascot characters, local idols, marathons, world heritage sites, and year-long historical drama television series. These are events that can be commonly found in many regions, but more often than not, no one really thought about why they are doing it. Concept is very important. If we just blindly follow ideas that were successful in other regions, our efforts and projects would quickly become passé.

For Mitoyo, we decided on the concept of conveying the culture of Sanuki udon noodles. And when we started thinking about what we could do with run-down *kominka*, the idea of creating a lodging where people could learn about udon noodles was born.

It is not ideas, but concepts that we should share. And if we are serious about this, I believe that different concepts will naturally emerge in each region.

To sum up, I have written down seven key factors that, I would say, helped to make the Mitoyo projects a success. I guess one could also say that this is the perspective we took when undertaking these projects that are now highly acclaimed by society.

Connections between the people who galvanize action

Hori: A lot of interesting words came up! We also paid a visit in person to meet two people, and asked them about their thoughts and feelings.

The first person we met was Kanako Harada, who jumped from a large corporation in Tokyo straight into the Mitoyo community and started Udon House, which offers udon-making experiences. The other person is Soichiro Imagawa, the third-generation owner of a local supermarket who was inspired by Ms. Harada's actions and success. He went on to start new brands such as "Soichiro Coffee" and "Soychiro Tofu."

The two met each other at a get-together for local business owners held at Udon House, and discovered that they both shared the vision of galvanizing people into action in the community. One thing led to another, and they started seeing new start-ups popping up one after another around them.

Miyase: Despite the fact that Ms. Harada and Mr. Imagawa are running different businesses, I felt the same level of passion from both of them when I interviewed them. Ms. Harada's determined attitude when she made the leap from Tokyo to Mitoyo galvanized Mr. Imagawa into action in Mitoyo as well. From their encounter came trust, which in turn led to the development of new businesses around them. I think it's not easy to connect with people and build a team that has the same views or goals like this, so how did you guys do it?

Furuta: We only get ourselves involved in "unaccountable" work. But this "unaccountable" means that we are not answerable to others or the community, and we push through with the work that we want to do. If you don't want to do it, you can just quit; if you are forced to do it, you're not going to hang in there for long. Conversations like this often come up. Even though I was the one who came up with the concept for Udon House, I told Ms. Harada that we would start the business on equal footing, not as employer and employee, and we both invested in Udon House.

Miyase: When I asked Mr. Imagawa how the locals viewed the activities

of Ms. Harada, who's not a local, he said that the local residents tend to focus on who is the center of the group when they get together by themselves, but this tendency is gone if an "outsider" is involved.
Furuta: There was a lot of commotion in the beginning. A young woman had moved here from Tokyo, and many wondered why and what she was going to do. But as we continued to meet Ms. Harada for drinks with others, we started to share our thoughts about what we each wanted to do and what we wanted for the community. In the end, Ms. Harada became the central figure that connected people in leadership positions in each municipality. It was she who led to the development of new relationships in the community.
Murakami: This is how second- and third-degree connections form in the community, and the relationships will become closer and deeper as they engage in more and more activities. And this was how a "mille-feuille-esque" multilayered community structure was created in Mitoyo as well—the key figures that form the core are surrounded by others who support it as well as those who are opposed to it.

"Opportunities to play a role" and "a place to belong"

Furuta: If the initiative/project is interesting, the relationships will continue for a long time, and will build the groundwork for creating new ventures or businesses again. Take, for instance, Silicon Valley: they have been able to attract people not because of the delicious gourmet food available there, but because the people know that in that environment, they may be able to create something new. In the same vein, Mitoyo has already become a place of choice for start-ups. And we believe that this is proof of the region's allure and its brands.

The key to building the community from this point is to create opportunities and a place to belong for human talent coming from within and beyond the region. No one wants to go for drinks or meetups where they don't get to say hi or introduce themselves, or where they don't know anyone else and don't feel welcome. We need to give them opportunities to do what they want to do, and a place where they can stay with peace of mind. It is important to build the community like this—by bringing people in from outside the region to connect to those who have awakened to new initiatives within the region, one by one, to form a serious and dedicated network.

Abe: As I was listening to you, I found myself amazed at how accurately you are able to judge people. Excellent management of human capital in a company actually requires us to carefully consider the positive aspects of a person and how to develop them and place them in the right positions. The same holds true for rural revitalization as well. It seems to me that the regions which have carefully placed people in the right positions are doing well. This is because both the people who made the judgments and the people who were selected strive to produce results or accomplishments that exceed expectations. I guess we could also say that the quality of the judgment determines the quality of accountability.

It's important to encourage people up

Fujisawa: I think a huge part of it is also because there are people like Ms. Harada and Mr. Imagawa, "super-committed" players, so to speak, who are extremely serious and dedicated. I myself am also working on a number of initiatives in the community, and I am aware that there are people who want to do something and have concrete ideas, but are unable to take the next step. So how can we find and nurture super-committed people?
Furuta: Mr. Imagawa wasn't all that serious at first, either. When an expert from outside the region came in, saying he wanted to be involved in Mitoyo, Mr. Imagawa left everything to that expert and ended up being at his complete disposal, and became depressed as a result. Right at that time, Ms. Harada came to Mitoyo and asked Mr. Imagawa, "What do you really want to do?" I think it was from that point that Mr. Imagawa's mindset changed.
Abe: It's important to encourage people up. If not, they may very well give up right away. Be it people from within or beyond the community, they need to fire each other up. In a nutshell, the key is to encourage people up!
Takemoto: In my case, there were many discouraged people in my area. Although this was also one of the elements that made it interesting, I've seen many situations where the locals are simply not passionate enough. Just as Ms. Fujisawa has said, we have not yet reached the level where the locals want to do a lot of things. But I am honestly not sure if this is a regional difference, or just individual differences, or a problem with the industries involved, or my own perceptive abilities.

Furuta: There are certainly regions where it is easier to bring about changes and regions where it is less so. If I may use the human body as an analogy, the surface of the body heals easily, but the internal organs take a much longer time. It may be that it is easier to start a new initiative in the service industry than in a primary industry. Maybe the reason why we could succeed in Mitoyo was because, unlike the neighboring towns of Kotohira and Kan-onji, which have established tourist attractions, there was nothing in Mitoyo. We were not influenced by any bizarre past success stories, and started off with a blank slate. Because Mitoyo was an overlooked place that did not attract anyone's attention, every small success gave us confidence: "Hey, we can do it too!" And I think that was actually a good thing for us.

Kamiyama: I also think that locals are generally discouraged. But telling them to "hang in there" is counterproductive. I think it is important to look for the good aspects, tell them about it, and praise them from the bottom of our hearts.

This may have been the same case for Mitoyo, too, but it is natural to have initial feelings of rejection and apprehension toward strangers hailing from outside their communities. Even if they do get praised by others, especially by people from Tokyo, they somehow won't believe it. In that sense, it was good to have many foreign tourists. The fact that they came all the way from overseas is very persuasive; furthermore, popularity with foreigners will attract even more foreigners. It was also good that we were able to effectively make use of the media for this. I believe the success we achieved restored faith and hope in the discouraged locals and created a cycle that will lead to the next one.

Financing and investment: a winning mindset in the community

Murakami: The reality is that rural revitalization is largely a result of the successes of individuals. In order to improve this situation and delineate a certain "model" (of rural revitalization), we need to factor out the problems from the "people" and "money" perspectives respectively. There have been many pointers given about people in what has been said so far. But what about money?

Furuta: I briefly touched on this in the beginning, but allow me to try to summarize the differences between financing and investment. Generally speaking, financing (loans) is based on past performance, while investments are planned based on future potential. Nowadays, we have crowdfunding and other forms of funding, and we are seeing changes, such as the ability to obtain donations for good ideas, even if one does not have any track record or experience.

Another change that we're seeing now is that the position of financing and investment may be reversed in the regions. When we are about to undertake a new project, we usually start by investing in it ourselves. But I am beginning to think that it is more effective to get financing on our own and start out with maximum risk.

Why? Because new challenges require decisive and rapid decision-making, and it takes a lot of time and effort to get people to understand something that is so new. Many fear that a vicious cycle will happen if things go south, and history has proven this so; we have seen many cases of locally financed facilities go bust or fail. If that's the case, things would move faster if individuals took on the risks instead. And indeed, that was what I had in mind when we started Udon House.

But for the Urashima Village project, we did a 180 and switched to investment. As I mentioned earlier, our efforts in Mitoyo were well on track and we could see a certain level of success. It is important that we do not keep our successes to ourselves, but share our experiences with everyone. And because we had also reached a point where it was possible to instill a winning mindset in the community, we decided to fund the project together after some discussion. Once people have that winning mindset, they will also have the courage and confidence to start something new.

From independent financing to mutual aid: mutual-aid financing

Murakami: I think we can describe the way the Urashima Village project was started as a sort of platform for fundraising. I think it is really significant that you were able to select from the community a group of people who resonated with the same vision, that is, the 11 members who invested in the project. There are not many examples in the region that have been able to create a cohesive group whose members do not differ in their degree of beliefs and passion. The logic of impartiality that municipalities have may deter a group of people whose hearts are equally passionate from standing out.

Furuta: I think there is also a problem with the fact that the financing system in regions is limited to "public aid = taxpayer's money," or "independent financing (self-help) = personal investment or financing." The mutual-aid system also simply doesn't work. What I have in mind is a "dedicated bank" for hardworking people in the community. It is not financing that you can't get if you don't have the means to pay it back, nor is it a subsidy that can only be obtained if certain conditions are met, but a form of mutual-aid financing in which people invest in or fund each other with the return of regional revitalization.

Takemoto: To me, it is a good thing if the project can be rolled out without government involvement through mutual-aid financing. But I often work on projects that deal with incredibly large common-property spaces such as forests and villages, so the number of the people involved inevitably increases. Even if some of the forerunners take the first step, it is still difficult to get the entire group to take action. When I think about what we can do to gradually make the majority of people involved, I start to think that it might be better to get the government more actively involved in some cases.

Fujisawa: Although many of my projects have received sponsorships from companies, I am having a hard time deciding where to look for returns.

Kado: It's really because, from a financial angle, there are debt and revenue issues. But from what I was told by a large corporation, they seem to also think that mutual aid-type services can benefit them if they can utilize the network. For example, trying to roll out an electronic payment service in a community or region would require a great deal of effort to go to each store and talk to them. But a shopping-district network would allow them to introduce the service all at once. I myself am the representative of a shopping-district association, and shopping districts are essentially mutual aid as well. I'd like to try various ways to raise funds through mutual-aid financing and create a context for it, too.

Abe: What I would like to do in advance is to create an infrastructure to get the people involved—as referred to connected minds—into structured finance. Tourism is probably a potential source of incoming funds; we can drop 50,000 yen for accommodations. Another source would be donations. For example, if we could create a framework for the people who are or want to become involved to do recurring or subscription donations, we would be able to obtain funds in a highly flexible manner that is easy to calculate. As a community, we can paint an image that the harder we work, the more funds we can amass. Recurring donations will also get the donors to be more deeply involved as members of the community.

The basic infrastructure model

Hori: This seems to be in the same vein as the Hometown Tax Program, but how deeply do they need to get involved with the community or region?

Abe: This is not something donors can expect to get in return, but simply a donation for the "experience" of being inspired by the community. All they have to do is to support them. In other words, it's like an entertainment expense to maintain relationships with the locals. For example, if I continue to do a recurring donation of 1,000 yen every month to Mitoyo, I know that I will be welcome when I visit the area and get to know various people. My staying experience would increase in value by 12,000 yen a year all at once. We could say that my return would be the "pleasure of support."

Furuta: It's interesting that you talked about recurring or subscription donations, because that is what I'd like to do in the future as well—to build a basic infrastructure for the community using a form of subscription donations. We are thinking of coming up with a model of mutual-aid financing—or maybe I could call it a new way of collecting taxes—where the locals can pool money together to cover public services in the community and use the surplus to further invest it back in the community.

The organization that will be handling the subscription donations project has its own power company, and provides manufacturing and transportation services. This would lower costs and also reduce living expenses. People who move to the area may very well be able to live on as little as 50,000 yen a month. This is a basic infrastructure model we are planning for the region.

And if we look a bit further ahead, there is the potential for establishing this model in multiple regions and connecting them to build a common infrastructure to serve the regions. This is the picture we are envisioning.

Hori: Mr. Furuta, do you have anything else to add before we end this session?

Furuta: I think IT will become very important in the future. The "IT" here does not refer to Information Technology, but Inspiration Technology—in other words, intuition.

There is so much information out there these days, and I think it's hard for people to make intuitive decisions. Not based on preferences, but on sensitivity that generates inspiration. It is important to train such sensitivity, and I think it is also important to create an environment for this.

This is also true of community well-being (residents' level of happiness in a community/region). How can we create an environment where people can intuitively feel happiness without being distracted by information? Recently, we have been having discussions from this perspective.

Session 2: Mr. Yoshiteru Takemoto's Case —Ama, Shimane Prefecture

A comeback story starting with Dozen High School: The shift from public aid to mutual aid, and then to "half civil servant, half X"

The town of Ama is located on Nakanoshima, one of the Oki Islands off the Shimane Peninsula. Facing a survival crisis with a population of just over 2,300, this small town made a comeback to see an increase in new residents, successfully attracting the younger generation by revitalizing its public high school as a cornerstone. This led to activities to develop industries that would enable graduates to return to the island, as well as private businesses that would support the town's future. Yoshiteru Takemoto of tobimushi Inc., who has been involved in the various initiatives on the island for 14 years, shares the full story.

For one to survive, all must survive

Takemoto: The regional revitalization projects that I am involved in are not about starting something pioneering, but rather, most of them are cases where the region has seriously addressed various issues in order to ensure its continued existence.

There are three key phrases for the Ama initiatives.

The first is: "For one to survive, all must survive." Dr. Takeshi Yoro once said this to me over a cup of tea at his home. According to Dr. Yoro, the treeline of the Himalaya Mountains is at 4,200 meters. The forests can grow at such a high altitude because many organisms had time to adapt as the Himalayas formed slowly in the process of the Indian subcontinent merging with the Eurasian continent long ago. Basically, few species fell behind and they were able to adapt as a whole. In other words, it's all or nothing. For one to survive, all must survive. I felt that this may be the same for rural areas.

The second is: "If you want to go fast, go alone. If you want to go far, go together." This is an African proverb, and it fits the Ama initiatives perfectly.

The last is Ama's catchphrase: "Nai mono wa nai" This has three

The town of Ama is located on Nakanoshima, one of the Oki Islands.

connotations: What we don't need is what we don't have; Everything that matters to us is here; Let's make what doesn't exist by ourselves.

Just what does this mean? Think about this as I explain some more.

Reform without self-sacrifice would not be supported

The Oki Islands comprise 180 islands, with the largest being called "Dogo." Nakanoshima (Ama) is one of the islands facing Dogo, along with Chiburijima and Nishinoshima. The magnificent islands as a whole are designated as a UNESCO Global Geopark. Of these, Nakanoshima is blessed with abundant spring water and flat land that is suitable for farming, making it an island of both farming and fishing with a food self-sufficiency rate of 140 percent.

Despite this abundant environment, the population decline was unstoppable. Efforts to build industries to attract people did not succeed, and economic collapse was impending. Ama's existence was indeed at a crisis point.

Amidst this, then-mayor Michio Yamauchi made the drastic decision in May 2002 to cut his own salary in half and reduce the salaries of town council members and town officials. The workers' union also volunteered to reduce their salaries. Based on the mayor's belief that reform without self-sacrifice would not be embraced, the town office took such initiatives before asking island residents for their cooperation, engaging the entire town in the fight for survival. Measures such as raising bus fares and returning subsidies were taken, and some of the funds secured by reducing salaries were directed to childcare-support policies.

The survival of the high school is the survival of the island

It was decided to use these scraped-together funds not just to extend life, but to go on the offensive. Thus, current mayor Kazuhiko Oe, who was section manager of the town office's industrial policy division at the time, took the lead in launching various initiatives, one of which was the Dozen High School *Miryokuka* (revitalization) Project started in 2007.

Dozen High School (officially called the Shimane Prefectural Oki-Dozen Senior High School) in Ama, the sole high school of the three Dozen islands, was already on the brink of closure. With a grim outlook for the island because of population decline and dying industries, student numbers dwindled as well. Some families were moving to the mainland before their children entered high school.

At this rate, the island would have no young people. Something had to be done. With this sense of urgency, current deputy mayor Misao Yoshimoto, who was sectional manager of education for the board of education at the time, set things in motion. Other adults responded, and the Oki-Dozen Senior High School Revitalization and Eternal Development Association was launched. Comprising the mayors, council chairpersons, superintendents of education, and junior high school principals of the three Dozen islands, as well as the PTA chairperson and the alumni association chairperson, the association members discussed how to preserve Dozen High School. I also participated as a member and sensed their conviction that saving the school would save the island.

And thus, the "island study" program was launched, with a call issued to students all over Japan. Teachers would provide intensive instruction to small groups, developing the potential of each individual. This went really well; the school has taken in over 150 students from outside the island to date. Some students had been living outside Japan through junior high, while some transferred from combined junior-

Fig. 1-2. Shift in number of newly-enrolled Dozen High School students

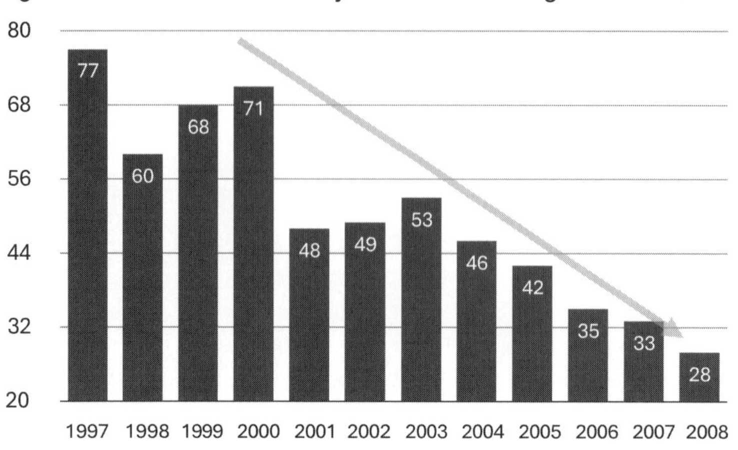

senior high schools.

A trinity of high school, boarding school, and public cram school

At the same time, we revitalized the dormitory and established a public cram school in collaboration with the school community. As we pushed reforms to integrate the three elements of high school, boarding school, and cram school, our reputation quickly spread by word of mouth.

As the number of students from outside the island increased, we launched a system of "island parents," similar to foster parents, to care for students who had just graduated junior high and come to an unfamiliar place.

Unlike student populations in the city, students' academic abilities and aspirations vary on the remote island, and there were no cram or preparatory schools. And so we established the "Oki Learning Center" public cram school to provide an appropriate curriculum for each student. The learning center has a "Dream Seminar" in which students can freely share their dreams. Discussions can become lively, going into the reasons for the dream and whether it is attainable on the island; sometimes an adult guest who embodies something similar to a student's dream will be invited, so the dream rapidly comes into focus. In a sense, the island is also a crucible of social issues where students can quickly gain social learning. The effects are showing in admissions test results,

Fig. 1-3. Shift in number of Dozen High School students

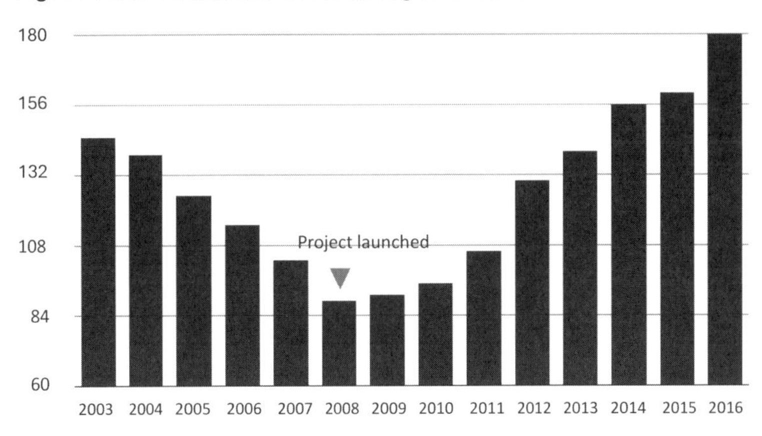

including university admissions office entrance exams.

About 80 percent of the high school's students attend this cram school. The teachers may someday be transferred away, but students will find familiar faces if they visit the cram school after graduating. This has become a wonderful third place in that sense.

Over 10 percent of the island's population are newcomers

How has the island changed as the result of these initiatives? One look at the graph (figure 1-3) will show you.

In 2008, a year after the project started, the number of students at Dozen High School hit bottom, then soared. It is said to be the only school in an underpopulated area that has student numbers increasing every year. The number of applicants from outside the island was 2.8 times greater than the number of students accepted, while local island student enrollment increased by 70 percent and the number of teachers doubled.

As the number of teachers increases, so do their families. As the connected population grew, the number of new residents also grew. In the 17 years from 2004 to 2021, 830 people (603 households) moved to the island. About half of those have stayed, resulting in over 10 percent of the population being newcomers.

August 2015 data showed the island to have a population of 2,352. This is 345 more than the population transition forecast of 2,007 people based on the national census. Projections for underpopulated regions are said to be very accurate, and if we consider the natural decline of the elderly population, more or less maintaining dynamic equilibrium for a decade is really amazing. It's truly an embodiment of the saying from the classic book *Hojoki*: "The river never stops flowing and the water is never the same."

A swirl of conflict: The "10-year ennui"

Hori: Thank you very much. The town of Ama no longer has a declining birthrate nor an aging population. It is amazing how policies centered on high-school reform were so effective.

Takemoto: But in the case of Ama, it would be hard to break down the elements of these initiatives and articulate what actions, and in what order, led to this result. I believe it is similar to what Dr. Yoro said, in that investing time for the overall initiative brought us to where we are today.

Hori: I also visited Ama and spoke with a number of people. My impression was that the regional initiatives expanded over the long period of 20 years, and the people of the island made discoveries little by little. An island parent who has taken in eight students said that the island study program changed the mindset of adults, and they no longer call people from outside the island *yosomono* (outsiders).

On the other hand, I also heard about the harshness of reality. Upon graduation, young people say "Thank you," and then they leave the island. There was also the strong dilemma of what to do about that.

Takemoto: That conflict definitely exists. As the students have a fulfilling time on the island and grow to have great potential for society, we do truly want them to flourish for the sake of Japan and the world. The students themselves also likely have ambitions for the outside world. But to be honest, we also wish they would stay to create the future of the region.

Murakami: What we learned from the example of Ama is that it's not as if someone first created the overall plan, drew up a detailed blueprint, and took action. Unlike the prior Mitoyo mutual-aid initiative, which was triggered by the independent financing for Udon House, this case started from a public-aid matter—a public high school—and evolved into mutual aid. However, it didn't roll out under the control of specific individuals, as if by the overall design of the municipality or a certain player. Rather, it changed gradually at the small epicenter of Dozen High School, while sparking discord and conflict in immediate human relations. That was what left an impression on me.

Fujisawa: If there was no blueprint, I'm very interested in how they created a system where children who felt uncomfortable studying in urban schools were accepted and flourished. I'm also curious as to how sustainability was attained even if discord and conflict were inevitable.

Takemoto: This is what we call the "10-year ennui." The school became popular, and a virtuous cycle seemed forthcoming, with an increase in students enrolling from outside the island, but then local parents started saying, "We want it to be a regular high school again." As it was the only senior high school for the three Dozen islands, other options weren't available to them. The school wanted students who came here despite having the option to attend a famous high school in Tokyo to harness the features of social learning and go on to prominent universities. However, we have learned that the parents on the island don't necessarily want that; they are looking for regular education.

At first, the school split the standard course into two: the Special Advancement course for those intending to advance to elite colleges, and the Regional Development course for those planning to work in that field. However, it turned out that most students had an interest in regional development, and an unexpectedly larger number of them chose the Regional Development course rather than the Special Advancement course. For that reason, we abolished the course system and integrated them both into a standard course, creating original learning opportunities within that. But then, some parents insisted that they wanted a "normal" standard course, while others felt it was important for Dozen High School to have its own character. It was impossible to reach a consensus. Without a blueprint, all we could do was move forward through trial and error. Now, in another try, we have organized a "normal" standard course as well as a Regional Co-Creation course intended for learning that delves into the region.

Murakami: That was right around when the first class of island study students were graduating from college and coming back to the island. Usually, most people would go on to take jobs at local banks or local government offices, but Ama didn't have enough jobs to accommodate

them. And of course, we thought, "Uh-oh."
Hori: And that's when a new development occurred. Let's have Mr. Takemoto explain.

The shift from full-blown public aid to "half civil servant, half X"

Takemoto: By the 10th year, we had executed many reforms under the excellent leadership of Mayor Yamauchi, and there were signs that the town would survive. This was the stage where the results of public aid were showing in macro figures. But on the other hand, it was becoming clear that the local economic cycle was being left behind.

This is a common phenomenon. When a tertiary industry such as education or tourism starts to take the spotlight, the primary industry that was originally central starts to lose vitality as a repercussion.

On the other hand, many students graduated and left Dozen High School, which had been revived despite many twists and turns, and some indicated a desire to return and contribute to the island out of gratitude.

The timing of these two coincided with the inauguration of Mayor Oe, and I think we were able to successfully change gears to take the next step.

One signature reform was "half civil servant, half X." Up to that point, we had reached a certain level of success under the "all civil servant, zero X" structure, in which the government handled all new challenges without relying on any nongovernmental organizations. However, when the island became filled with motivated youth, including alumni, the government alone could not provide them with sufficient opportunities. We had to create momentum for private businesses. In short, the school served as a center for developing human resources, and we increased alumni and the connected population by operating this steadily. To have this essential workforce drive the energy of the island forward, we adopted the "half civil servant, half X" system.

From public aid to mutual aid with the Future Co-Creation Fund

What specific measures did we take? One was the establishment of the Future Co-Creation Fund. We decided to allocate a quarter of the income from the Hometown Tax Program to finance this fund, which would invest in private business ventures that will support the future.

Fig. 1-4. "Half civil servant, half X" system

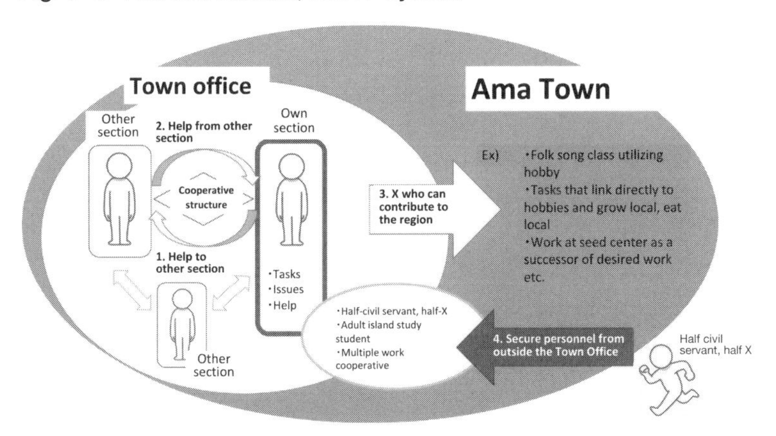

For example, if someone wanted to open a bread bakery on the island, the market scale would be insufficient for securing financing from a banking institution, yet it would also be difficult to obtain a direct public subsidy. The fund was created to support business launches like this, which pose a challenge for involvement in terms of both market and public policy.

These are public funds, as we are using the Hometown Tax, but by placing them in a private fund and separating them from the general account, we become able to use them for business without going through the council approval process. This involved very difficult decision-making, but we overcame the challenges to achieve this.

For this reason, we are very careful to ensure the legitimacy and fairness of the businesses we support. We not only discuss the business plans multiple times with investment applicants, we also require them to make a presentation at the Ama Town Industry Culture Festival, which is a major festival on the island. For this, they have the huge responsibility of properly explaining how the business will benefit the future of the island and convincing island residents, who mostly know each other.

Following the launch of these initiatives, Hometown Tax donations soared from 40 million yen to 120 million yen in one year, and to 215 million yen in two years. One-fourth of that is allocated to the Future

Co-Creation Fund. The groundwork is being steadily laid to use these effectively as mutual-aid funds.

Fostering the next generation of leaders for the island

The next initiative was the Multiple Jobs Cooperative. This permits migrants to take on multiple jobs at first, so that they do not have to leave the island because their employment wasn't suitable, despite wanting to stay.

Island societies have extremely low human-resource mobility. The risk is huge for both employer and employee when hiring someone who has just arrived from outside the island. Also, there are jobs with fluctuating needs, such as in the primary industry, which has an off season and a peak season. Accordingly, a cooperative was formed by multiple businesses to enable people to work different jobs every quarter.

If a person finds a workplace that suits them during this process, they can leave the cooperative and sign a separate employment contract. Another possibility is to maintain multiple jobs. In a small island society with limited employment opportunities, this is an excellent initiative to optimize them.

I would also like to explain the "half civil servant, half X" ordinance. Simply put, this is a system where civil servants can also engage in work in the private sector. In rural areas, competent workers are often drawn to government offices. The question was how that talent could be shared with local private companies. Amid momentum for workstyle reforms and endorsement of side or second jobs, couldn't a system be developed so civil servants don't violate the obligation to devote themselves to their duties? This was achieved through steps such as amending the ordinance.

Under this system, when a civil servant finds certain work (X) that can contribute to the region and is authorized by the mayor based on impartial criteria, they may either take a temporary leave from their civil-servant duties or engage in X as a side job.

These initiatives were not all planned from the start; they are simply the result of focusing on immediate matters, one by one. But looking back now, we can see that they are connected at a fundamental level by the message of "fostering the next generation of leaders for the island." The "half civil servant, half X," the Future Co-Creation Fund, the

Multiple Jobs Cooperative—these were all devised for the sake of developing human resources for the future. Ultimately, all of these efforts aim to leave the sameone result. I believe this is what the region is all about.

The "cell membrane" (community) and the "nucleus" (people) that embrace the region

Abe: This is intriguing. Within the closed environment of a remote island, there was no option but for the local government to take the initiative—ownership in a sense—to move things forward. Using an organism as a metaphor, just as the nucleus can realize the need for its existence as it is in its own territory covered by the cell membrane, the presence of the powerful "cell membrane" (community) of Ama made it possible to successfully engage the "nucleus" (people) with a sense of relevance. That is my impression.

But in most regions, it is not typical for people to take ownership as individuals, so there may be a need to discuss how to apply this case to such regions. The town of Ama is commendable in that it had an environment for taking ownership in the first place, and that systems such as the "half civil servant, half X" and Multiple Jobs Cooperative were developed to further amplify that. Requiring a presentation to island residents to qualify for investment is also an excellent move, as the party making the statements becomes obligated to take ownership.

The approach may be the opposite of that taken in Mitoyo, but it seems that two conditions are requisite to sparking change: taking ownership of a field and having autonomy and aspiration toward the business.

Furuta: This may be the same thing as the "opportunities to play roles and places to belong" we refer to. Not having roles to play gives rise to a sense of alienation, while a place to belong on its own makes people passive. It is crucial to have both, and in concert, with the right balance. In Mitoyo, we had the story of starting by developing roles to play and then finding the places to belong. In Ama, developing the places to belong was first, and the roles to play came later. It's definitely the difference of starting with independent financing versus public aid, but in the sense that both "roles to play" and "places to belong" were ultimately created, I think it was the same.

There was mention of a sense of crisis with young people leaving the island, but I actually think there's nothing wrong with that. I guess it depends on what you consider "sustainable," but can't having nothing beyond high school be considered an attractive and vital element of the island?

Takemoto: There are definitely people leaving, but there are newcomers as well, and the island demographics are actually improving.

Kado: Considering that the island country of Japan is also enclosed in a "cell membrane," Ama and its various issues are a microcosm of Japan. I feel that this case showed us how much Japan can change if galvanized in the right way. You could call this "nucleus and cell membrane" or "function and place." The question is how to combine these to enrich and bring happiness to people's lives. I have a feeling there are potential hints in the initiatives of Ama that can guide the island nation of Japan to the essence of well-being.

Regional management systems centered around education

Abe: An island parent interviewed by Mr. Hori said that they found out only recently that the word *yosomono* (outsiders) is discriminatory. I felt this was an important realization. As this person had likely taken in several children, they were made vulnerable in many senses and took on risks, but the students observed this and matured as well. I think that this entails the most important device in education today—having contact with an adult who is in a vulnerable state. Many schoolteachers have not been exposed to society or risk, so their instruction feels like preaching and doesn't resonate with the children. They may be studying, but they don't learn. The teachers at Ama are always vulnerable, and this has resulted in the best educational environment. That is my impression.

Takemoto: It is definitely an island that is vulnerable and full of pressure. [Laughs]

Kamiyama: It was a bit too perfect. I thought it was too good to be true. [Laughs] Preserving the sole high school, with young people arriving from outside to live there. Simply having these children here in the region raises hope for the future. That in itself greatly changes the overall circumstances. As I said last time, most rural areas are full of people who have no hope for the future. While they are aware that something has to be done for the future, it becomes an issue of, "So who's going to do it?"

Maki: Mr. Takemoto's earlier words, "'Fostering the next generation of leaders' is a message that connects all" struck a chord with me. Apparently there was no comprehensive blueprint from the start, and although they each seem self-contained, they actually are mutually relevant. I do think the plans were carefully thought out.

Also, the Ama board of education has a business-planning function for some reason. This is not common for a board of education. Ama had a regional management system centered around education. I think this was actually a vital factor. The town of Ama today has solid industrial policies to create employment, business, and demand; these were driven by an intent to foster the next generation and a sense of crisis that the people they had fostered had nowhere to go. This development is suggestive of Ama's next phase.

The deep democracy of a society where everyone is assertive

Hori: The process of shifting from "developing human resources" to "developing industry" seemed to trigger a range of challenges and trial and error for the locals. When we visited the town office to research "half civil servant, half X," the executives told us, "We started out as a youth organization, so our ties are strong. However, the youth organization is now gone. The current issue is determining who will take over after us." In the past, public officials would negotiate with landowners to carry out public construction. As they communicated and were shouted at, they became really familiar with the region and were able to conceive lots of ideas. But there is no longer public construction. That is why there is need for a system like the "half civil servant, half X" to integrate the private sector frontlines. They were saying that they wanted to develop the next network in place of the youth organization.

Fujisawa: In the video interview by Mr. Hori, the manager of the Entô hotel was saying it was perpetual self-questioning, asking, "Whose sake does this island exist for?" Those words stayed with me. I think it's similar to "Who does this country of Japan belong to?" and I sensed that the ideal of what could be called "deep democracy" was budding here.

Takemoto: It just so happens that self-government and democracy is a hidden theme of this case. I consider "society" to be similar to "the public," and that the public is a collective of what seems to be "We the People." The town of Ama clearly has this "cell membrane" that is We the People. In other words, the consciousness of "We the People as the island" is very strong. Within that, there are also smaller cell membranes in units of villages.

For example, the tug-of-war competition, which is a traditional island event, may be a symbol of this. This is a battle they can't lose, with everyone putting the pride of their district on the line. I was enlisted two days after I moved to the island. An older lady I didn't know approached and upbraided me, saying, "The way you're pulling is no good. You have to apply your weight like this!" And so you have the district as the small cell membrane and the island as a large cell membrane, and the whole forms one society.

It is said that 150 people is the limit for a society to live without rules by being aware of each other. So when that number is exceeded, communities need social rules to a certain extent. After passing 150, what is the maximum population that could be considered "We the People"? Referencing the words of luminaries such as Aristotle, Thomas Jefferson, and Jean-Jacques Rousseau, that would likely be in the range of 5,000 to 20,000 people. The population of Ama is 2,300. Ama, which has established a community on a moderate scale, has been able to embody the importance of regions having a certain level of closeness and building upon discussions, as Mr. Murakami mentioned.

And so what defines a society that is functioning as a community? I believe that this would be a society where individuals recognize the relevance of the sustainability and non-sustainability of various matters. Simply achieving carbon neutrality does not automatically make a society sustainable.

Just how will individuals respond to that sustainability? Obviously, there will be both support and opposition, as well as action and reaction. I think this is fine. There is no need at all for everyone to face the same direction at the same time. What's important is that a range of opinions are integrated on the premise that people are taking ownership.

And so I think that a functioning community is a group of people who can respond to matters around them, and that this "response ability" would be the equivalent of the word "responsibility."

A society that can respond to the various matters of the region—in other words, this is a group of citizens who take responsibility for the future. In contrast, when a society has become unable to engage in dialogue with responsible citizens, democracy can be considered dead. It transforms to a large mass of apathy. In that sense, I believe that Ama is taking responsibility toward the future.

One last thing I'd like to add is that Ama is a community of "*Minna-de-shabaru*." I love this commonsense slogan, which is from the town administration under Mayor Oe. While "*minna*" means everyone, "*shabaru*" in the dialect of Ama means "to pull strongly," so "*Minna-de-shabaru*" means "Everyone pulling together." At the same time, this also affirms "*Minna-deshabaru,*" which means "everyone is assertive." The assertions of individuals are not left hanging, because there is an understanding that everyone will pull together. I think it's very typical of Ama to use this as a slogan.

The message of "For one to survive, all must survive" is also engrained in this; the town of Ama is practicing "*Minna-de-shabaru*" to sustainably preserve the entire region.

Seeking rebuilding of closeness for rural revitalization

Miyase: Unfortunately, I wasn't able to visit for research, but listening to the words of those in the video, I was given the impression that they sense the challenge of clearly differentiating between public and mutual aid. There is some warmth and humanity in public aid, and a certain scale or scope seems to be needed for a close-knit community to function, as Mr. Takemoto said.

Murakami: I don't know if this applies to other regions as well, but in the case of Ama, developing "people" in the easily discernible form of Dozen High School resulted in the successful trial-and-error undertaking of rebuilding a system similar to mutual aid that was on the verge of being lost.

A common factor of faltering rural revitalization is that people are attempting to simply duplicate the system of a successful region, even though the region to be revitalized does not yet have sufficient closeness. It is tricky to apply best practices horizontally unless the region has built close human relationships by sharing a certain sentiment as the result of experiencing conflict and confrontation. In other words, I think the rebuilding of closeness is important.

Furuta: Outsiders assess this type of initiative as being excellent upon seeing it for the first time when it has already reached a certain level of completion. But what's important is its starting point, which can affect the overall image of the initiative. For example, when a makeup artist changes a single eye line, this sets off changes in the hairstyle, accessories, clothing, and even behavior. It is the same for Ama. Changing education became the starting point for overall change.

Abe: If we take a slightly different perspective, the content of this education that served as the starting point for regional reform endorses innovation. The aim is to foster innovators. That is why the desirable human resources that were cultivated there will aim for the next stage and leave "We the People" [of the island] to seek something different from others. It may actually make sense for them not to return to the island. After all, that means they want to make something that doesn't exist here. But if they return to the island, they are expected to take responsibility. This leads me to think that innovation and democracy inherently don't match. A contradiction cannot be avoided.

Murakami: There is the aspect of that paradox spawning power.

Kamiyama: I think that Ama's initiatives embody the ideal status of Destination Management Organization (DMO). However, there is no formal DMO in Ama. There are often cases of regional-coordination DMO failing because of the presence of too many leaders, but with a municipality the size of Ama, it is easier to establish democracy.

Overcoming divide to reach innovation

Hori: Whether you are inside or outside, anyone who has passed through Ama even once can engage with the island anytime from anywhere. There were people who said they wanted to create connected minds like this.

Kamiyama: One day, there may be a shift to engagement in the form of "kind of living there, kind of not living there," like a metaverse.

Kado: Humans habitually seek identity, so I think that no matter where they go, they will create a "cell membrane" and feel secure as long as they are inside it. Ama's method of embracing people like that was successful. However, it is meaningless unless the story born from a cell membrane doesn't connect to the next story. To that end, there is a need to integrate people with diverse values.

Abe: As each individual has their own role in the community, dispersal may be part of their development process, while at times, they may come back. Amid this gradational environment, general strategies are formed while increasing the number of connected minds. That may just be the only way.

Maki: Regions have the unspoken rule of "functioning to avoid conflict." This could be considered deference-type consensus-building in a mutual-monitoring society. Matters are well coordinated, so there is no need for battle in elections. But when new people move in, the ecosystem is disturbed and division is manifested. My sense is that this triggers bidirectional processes, leading to tolerant societies with a culture of mutual acceptance of diversity.

Some may feel stress from getting entangled, but it is an imperative step to enhancing closeness. I believe that there is also meaning in manifesting division.

Takemoto: It's true that in Ama, the intent to expose division has become stronger over the past few years. When people need to escape from crisis, division is irrelevant, and there is no option other than to align and run in the same direction. But now that we have entered the next phase, people are calling for acceptance of opposition. As a range of approaches and initiatives emerge and the gradation becomes wider, the question is what will happen now. The real test for Ama lies ahead.

Maki: Back to what Mr. Abe was saying, an innovative workforce will remain after overcoming the manifested division. That is what the future looks like.

Fujisawa: I agree. I've been supporting corporate innovation programs, and there really is a methodology that starts by manifesting division in a community and then sparking innovation as a result.

Hori: In the exhibition room at Entô hotel, which was mentioned earlier, you will see a sign that says, "The continent broke up, a volcano erupted to form a caldera, and now there are three islands. These are the Oki Islands." According to the latest research, the separate islands will merge into one in 250 million years. The people of Ama may have actually been subconsciously aware of the themes we discussed here today.

Session 3: Mr. Yasuhiro Kamiyama's Case —Institutionalizing Private Lodging

How can we create market-creating industry associations? The art of lobbying: trusting and working with politics and government

Yasuhiro Kamiyama from Hyakusenrenma, Inc. was instrumental in getting the Private Lodging Business Act passed from the perspective of the private sector. He brilliantly reconciled the interests of politicians, government, and operators regarding private lodging with his excellent communication skills, and succeeded in turning it into a business for the local community. Mr. Kamiyama speaks of being driven by an "inexplicable sense of mission." He also shares his art of bringing diverse players on board—an inevitable part of driving rural revitalization. In this conference, the trailblazers will exchange their opinions on this amazing expertise.

Establishing rules to leverage private lodging in the community

Hori: I think there was a time when the term "illegal private lodging" was all over the news. In the mid-2000s, "private lodging," in which private homes and facilities are rented out as lodgings via the Internet, became popular in Japan. However, most of them were illegal operations, and from the point of view of existing hotels and inns, they were obstructing business. It was also problematic since it came with issues like trouble with residents in the vicinity, crime prevention, disaster prevention, and sanitary problems. And it was Mr. Kamiyama who worked hard to establish new rules to leverage private lodging.

Kamiyama: Actually, I was the one who coined the term "*yami* [illegal] private lodging." It all started when I told a newspaper reporter about it. By right, private lodging has the potential to create new tourism services and experiences for the region. If we can establish rules to properly control it, we should be able to open up a new market. With that in mind, we decided to take it up with the government.

Murakami: If I may add on a bit more, lodging facilities in Japan have business permits under the Hotel Business Act. This act, which is under the jurisdiction of the Ministry of Health, Labor and Welfare, was originally created for public health purposes, such as countermeasures against infectious diseases. The reason why all the lodging businesses have to keep a guest register is because if an infection occurs in a place where a large, indefinite number of people gather, the source of the infection must be identified. On top of these sanitary regulations, there are several other binding obligations, including those under the Fire Service Act. This is why there's also the aspect of regulating the new entrants into the lodging facilities market.

One of the reasons private lodgings have become a problem is because the number of facilities operating without complying with these regulations has increased significantly. And if left unchecked, the legality of private lodgings will be subject to question, and may even smother new industries in their infancy. It would also hurt existing inns and hotels that comply with regulations and disrupt getting guests through the regular channels. This was the situation we were facing back then.

Kamiyama: Yes, that's exactly right. This led to several measures, including the designation of National Strategic Special Zones and the establishment of "Special Zone Private Lodging" in December 2013 as a special exception to the Hotel Business Act. Based on this, we finally managed to get the government to enact the Private Lodging Business Act in 2017. This Private Lodging Business Act is under the jurisdiction of the Japan Tourism Agency (an agency of the Ministry of Land, Infrastructure, Transport and Tourism). And it has been a long road to get here.

"What we are doing is supposed to be right, but why are we being told it is wrong?" I can understand the grumblings of the private lodging operators and the feelings of the online travel agents (OTAs) that are their middlemen. But since we are living in a country where no one is above the law, then everyone must follow the existing rules. And if necessary, is it wrong to make new rules?

This was how we talked to the business operators. Since we had to talk to those in the inn industry, people from various associations and unions also came to the table. I spoke to all relevant national government officials and members of the National Diet (Japanese parliament) on an individual basis, and got various opinions and thoughts. It is important to keep in step with new ideas such as the sharing economy. In the end, I even had the House of Councilors, which was in the process of deliberating the bill, invite me as an unsworn witness so that I could underscore the need for a new act.

I went in said, "Ladies and gentlemen! Please listen to this one last request of mine! Please absolutely refrain from issuing agent's licenses to lodging-business operators that market illegal properties!"

And to the operators who provided the so-called illegal private lodging, I implored them to do it under the legally recognized rules so that the various stakeholders or people involved would have a better understanding of the situation, which should lead to further relaxation of restrictions.

The first step is coordinating interests to resolve issues one by one

Murakami: Under the Fire Service Act, sanitary laws, and other regulations, private lodging facilities using *kominka* (old traditional Japanese houses) and others would have to fork out a large investment if they were to comply with the regulations. On top of this, to follow the regulations they would need to set up a reception desk/counter as well as keep employees in a house that's to be rented out in its entirety. This would inevitably put a damper on the rental experience, making it lose its meaning for guests who want to rent private lodgings. This is why they started off by making use of Special Zones—because it would be easier for them to initiate bold regulations reforms.

Kamiyama: We approached the Cabinet Office through my company, Hyakusenrenma, with the plan of kickstarting the process by making use of the National Strategic Special Zones. But it proved to be difficult to find local governments that were willing to participate. I got the impression that they wanted to avoid getting into sticky situations. In the end, the first few to agree to give this new idea a try were Ota Ward in Tokyo and the city of Osaka in Osaka Prefecture.

Murakami: You managed to get the Private Lodging Business Act enacted by coordinating the interests of various stakeholders and parties, but you still faced some challenges, if I recall correctly.

Kamiyama: Yes. According to this act, the people behind the business of private home rentals (i.e., private lodging) can be simply defined as the hosts who operate the lodging business, and then there are also the agents who provide information and handle reservations between the operators and guests. The OTAs that I mentioned earlier would serve in this capacity. This would also include administrators that are entrusted with the management of private lodgings when there are just a few rooms or when the owners of the homes are not around.

The first challenge that we had was that only real-estate business operators were allowed to be such administrators. For example, even if there's an operator of a another private lodging facility, hotel, or inn in the vicinity who is willing to look after the private lodging facility in question, they would not be allowed to do so simply because they are not real-estate business operators. Allowing localized travel agents and Destination Management Organizations to take on the role of administrators should lead to the generation of new businesses in the community or region.

The second challenge was that even after becoming a host, one can only rent out their property for up to 180 days in a year. This means they can only run the business for half a year! That was the biggest obstacle we faced. But despite these challenges, we continued to work on this project. We thought that we would eventually be able to get everyone to understand better if we stuck to this framework for now, and that it would pave the way for further relaxation of the restrictions.

Willingness to integrate into the community

Hori: During the course of establishing these rules, Mr. Kamiyama also worked hard to establish industry associations for private lodging operators. I believe they are the Japan Association of Vacation Rentals (JAVR) and the Japan Countryside Stay Association (JPCSA). This has fortunately led to the creation of a new framework in the realm of private lodging, which had been developing in an unregulated manner, and eliminated illegitimate, illegal private lodgings.

Kamiyama: The real question here, though, was whether or not they were willing to integrate into the community. These two organizations were created so that willing operators could openly disclose their intentions.

The JAVR is a group of local and international OTAs that handle private lodging. It is the only organization in which almost all the OTAs operating in Japan participate, and the majority of inbound tourists visited Japan via the listed members' websites. Although the number of inbound tourists took a temporary dive due to the COVID-19 pandemic, the expansion of the inbound tourism market is part of the national strategy, and the activities of these members will generate very significant positive impacts for the country's economy. And if that is the case, someone will have to bring them together into a group.

The Japan Countryside Stay Association is a public-private partnership that promotes "countryside stays" in farms; participating members include operators of private lodgings and countryside stays, local governments, and large corporations. Instead of just offering homestays in farmhouses, the objective of Japan Countryside Stay Association is to also make good use of local properties that are idle (such as vacant farms, abandoned houses and vacation homes) so as to offer entire-property rentals in all types of vacant standalone dwellings like *kominka*, vacation homes, castles, and former samurai residences. To be more specific, our work is to help those who are already doing this but not getting much in the way of good results or guests. This association is helmed by a number of prominent members including Shinji Hirai, governor of Tottori Prefecture, as our first chairperson; Masao Uchibori, governor of Fukushima Prefecture, as our current chairperson; and Yoshitsugu Minagawa, former vice-minister of the Ministry of Agriculture, Forestry and Fisheries, as president. We have also formed a policy committee to make proposals to the government.

I have been asked to run these two associations as representative director, and part of my work is to facilitate the exchange of opinions with various stakeholders and parties like the Parliamentary Association for the Promotion of Farm Stays (countryside stays) and the Parliamentary Association for the Promotion of Workcations ("working vacations"). It is also important to have the understanding of National Diet members in order to proceed with what we want to do in a concrete manner. Just the other day, I attended a meeting of the Liberal Democratic Party and proposed the idea of using private lodgings to help realize the national vision of Digital Garden City Nation. I also added that we are a properly established and legal association. [Laughs]

We also visit private lodgings in regions in person. There is one called Misugi Resort in the city of Tsu in Mie Prefecture. Here, local inns and hotels, which would normally be expected to oppose private lodging, are actually playing a leading role in promoting the use of abandoned houses in the neighborhood as "detached guesthouses." These detached guesthouses can serve as a type of private lodging for guests who want to stay for a long period of time. And if we are to further roll out this concept in other communities or regions, we need to relax the restrictions on administrators that I mentioned earlier.

Albergo Diffuso: A measure to stimulate tourism

Another recent initiative that I'd like to touch on is called Albergo Diffuso (AD). It refers to a framework of decentralized hotels (*albergo diffuso* in Italian), also referred to as "community-wide hotels," scattered over the entire region.

Hori: We know that it is difficult to personally manage each and every idle private home and facility in the tourist areas. AD is actually a town-wide initiative in which one core facility is established to serve as the reception center, which will in turn lead to the lateral development of restaurants and lodging facilities in abandoned houses. I heard that it started in Italy about 40 years ago as a solution to the problem of abandoned houses and a declining population.

Kamiyama: Yes, that's right. We would like tourists or guests to stay at places like *kominka* hotels and experience living like a local by trying local foods and sightseeing. This is a type of stay that straddles somewhere between "staying" and "living," where one can get a feel for the culture and history of the town and live like a local resident for a short period of time.

In the case of Japan, I think the old post-station towns and castle towns would be suitable to function as ADs. In fact, there is an old post-station town in western Japan called Yakage, in Okayama Prefecture. In this area is a *kominka* inn called Yakage-ya Inn and Suites that has been accredited as Japan's first AD, and functions as the core base of the town. Albergo Diffuso Internazionale (ADI) is an organization that promotes the spread of the concept and activities around the world, and accredits any facility as an AD if they can meet the ten criteria set forth by ADI.

In recent years, local governments have also started rolling out a framework appropriately called "Albergo Diffuso Town (ADT)" to accredit such facilities. This also includes another concept called *ospitalità diffusa* (OD), or scattered hospitality, along with ADs. Fundamentally, it is the same as AD, but OD covers a wider area—about the size of a district—with facilities spread out over a radius of approximately one kilometer from the core base. (For ADs, it is within a radius of 200 meters.)

The person who first came up with the concept of *albergo diffuso* was Giancarlo Dall'Ara, now the president of ADI. He has been visiting Japan a lot lately, and commented, "Because the villages in Japan cover a vast expanse of area, it would be good to think of the range in terms of kilometers and have separate operators work loosely with each other." That is OD.

ADI seems to hope to accredit about a hundred ADTs in eastern Asia, especially Japan, within the next five years. This has also led to the establishment of a Estremo Oriente (Far East) branch in July 2022 under the direct management of ADI in Japan for this purpose. This accreditation will bring many tourists from all over the world; very straightforward and easy to understand. This is just the perfect initiative to prepare for the next wave of inbound tourism.

Turning vacation home areas, castle towns, and abandoned farmland into tourism assets

With that in mind, the town of Zao in Miyagi Prefecture has rolled out an initiative that also taps into the idea of *albergo diffuso* in an attempt to breathe new life into its once popular vacation home areas by transforming them into lodging facilities after the Private Lodging Business Act was enforced. There were many decrepit and run-down vacation homes built around 1965–1975 that looked like they were going nowhere. My company and N Corporation K.K., a real-estate company, decided to help drive this project together, and after many discussions with the locals, we finally came up with the idea of repurposing them as lodging facilities.

We started with just one vacation home. Now we have expanded it to about 40 vacation homes in the same region. Our original target were foreign tourists, but they all disappeared with the COVID pandemic. Surprisingly, the number of domestic tourists staying for long periods of time has increased dramatically, and the occupancy rate is 70 percent at worst, and 90 percent at best.

Based on this case study, the city of Hachimantai in Iwate Prefecture is also in the process of transforming countless vacation-home areas into lodging facilities. Since their geographical range is a little bigger, we are thinking of getting the entire city accredited as the first *ospitalità diffusa* in the world. But they are somewhat lacking in the gourmet aspect, and we had to work on this. We therefore created a facility called Northern Grande Hachimantai that offers guests the opportunity to try out local cuisine as well as serve as a core reception center. Also, here with us today is Ms. Daibo, who took on the risks herself and worked day and night for this facility.

Hori: I actually visited the restaurant for an interview and thought that it was extremely well designed, with offerings of plenty of local ingredients with local flavor and a stylish ambience that made me want to eat there. It is also a base for providing new value to the community.

Kamiyama: To add on to it, there has been geothermal power generation here for a long time. Their access to sustainable energy gave rise to the momentum to brand the region as a sustainable community.

Another case study is the "castle stay" in the city of Hirado in Nagasaki Prefecture. We worked with a company promoting the Noroshi project to turn Hirado Castle into a lodging facility, as we felt that it would be a pity if the castle was only used as a museum of history and culture. Because many of the castles in Japan today were actually restored around 1960, local governments now have to decide whether to reinforce them against earthquakes, rebuild them, or just give up on them.

As this "Kaiju Yagura Hirado Castle Stay" is the first castle where guests can stay in Japan, we started out by slapping on a high price tag of a million yen each night, targeting the upper class. We also limited it so that only one group was allowed to be the special "castle lord" for each day. We initially prepared an accommodation plan priced at one million yen, thinking that we should offer this much services to the "castle lord." Then, strangely enough, we somehow managed to craft a less expensive plan at 500,000 yen. This made me think that it might be a good idea to start out with expensive products or services as a way to gain empirical knowledge of the region.

We are also working on a project to offer stays in castle towns in the Obi district of Nichinan, in Miyazaki Prefecture. The framework of our plan is to combine the Sakamoto Rice Terraces, one of the top 100

Hirado Castle is the first castle in Japan to allow guests to stay overnight.

terraced rice fields of Japan, with the quaint and historic townscape to form an *ospitalità diffusa* that would offer countryside stays. Another case study is the town of Osato in Miyagi Prefecture. There are many abandoned farmlands over there, and our plan is to create a model where these farmlands are turned into *kleingartens* ("allotted gardens") where guests can rent a small plot of land and stay there to farm for longer periods.

Along with these discussions, I also had the opportunity to talk about the importance of utilizing local cultural assets and idle properties in a tourism-related task force during the years when Mr. Yoshihide Suga was serving as the chief cabinet secretary.

Driven by an inexplicable sense of mission

Abe: I am just so, very wowed by what you have done and achieved. You've gone so far as to deal with politics and even create associations to legitimize a new category of travel. The harsh reality is that travel-related industries are under the strict purview of laws and regulations, and it is difficult to make a profit unless you are a very large company. And if you work for an association, you can't really put yourself out there to make a profit, and the company probably won't get much return despite all the efforts that you have put into it. This is why I think what you've done is so amazing and commendable; truly a godsend for the Japanese society.

Fujisawa: My thoughts exactly. You have created industry associations as a window for the operators to be able to communicate with the government and the administrative agencies. I think that is wonderful. When someone wants to start an initiative, they do not know where to go first and what to ask, and there are bound to be difficulties in communications between the government and regions.

What you've said about *albergo diffuso* was also extremely interesting, and I listened to it thinking that this will probably turn into a trend in the future. But on the other hand, I also wondered who in the region would be the one to bring this up. When I tried to start an initiative in a certain region, I was told by the locals that it was fine if they didn't make any extra money. It felt like they wanted to get their hands on the grants right away, but they were not keen to start anything new. . . . I still have memories of walking door to door with my head bowed, as I grappled with the difficulties of the community's unique herd mentality.

Hori: I had the pleasure of following Mr. Kamiyama on his trip to Hachimantai and covered more on this. Mr. Kamiyama was very attentive and asked so many people, from the decision-makers to those who draw up the blueprints, questions like: Where are the issues in the community? What are the needs on the ground? What sorts of regulations or relaxation of regulations are needed? How is the cash flow? And he has made many such trips.

He also visited the Liberal Democratic Party Headquarters and politely proposed to a number of key people, one by one, that they help him out with his requests. Mr. Kamiyama has striven to be a partner who speaks for the people of the community, instead for of any particular group or his own company. His selfless attitude has left an indelible impression on both me and Ms. Miyase.

Abe: Mr. Kamiyama has once described himself as "a politician in the private sector," and he couldn't have been more accurate. And it also seems that you have to engage in a lot of intense tactics.

Kamiyama: Well, not really. When I saw myself in Mr. Hori's interview footage, I thought to myself, "What in the world does this old geezer

want to do?" [Laughs] That's not a business project or operation, no matter how you slice it. Then why am I doing this? When I talked with people in the community about this or that, they tended to throw in the towel right at the beginning, saying, "It'd be great if that really happened, but nothing's gonna change anyway." And then I felt, if that's the case, I'm gonna go all the way to Tokyo and talk about this. I guess I'm doing this because somewhere in me, there's this inexplicable sense of mission that's driving me forward.
Hori: I'll never forget the sight of him running around the Diet members' office building and ministries with a backpack on his back. So, Ms. Fujisawa, I guess you could say that the person who will bring this up is the person sitting right before us, Mr. Kamiyama.
Fujisawa: That is so true. It was Mr. Kamiyama who visited the locals in person to find out the missing pieces of the puzzle, and it was also Mr. Kamiyama who tried to fill in the missing parts. But it also made me wonder, can someone else fill Mr. Kamiyama's shoes? And how can we pass on this know-how and knowledge?
Hori: That's an interesting point that we should keep in mind as we proceed with the discussion.

Wearing both the hats of a lobbyist and a business player

Maki: That was really informative and enlightening. Personally, I was somewhat averse to organizations like industry associations to begin with, and was convinced that they were simply a group of people who would be opposed to trying anything new. So it came as a huge surprise to me that a "market-creating industry association" like the JAVR would ever even be created. I am definitely going to take a page from Mr. Kamiyama's book.

In my case, I have been gradually producing small projects in areas such as forestry, agriculture, employment support, caregiving, lodging, and food and beverage services in three locations—the village of Nishiawakura in Okayama Prefecture, the town of Atsuma in Hokkaido, and the city of Takashima in Shiga Prefecture—and have steadily continued to achieve results, one by one, which are unconnected to each other. I'm currently trying to find a way to one day integrate them into a framework that will bring about a structural change in the local economy, and I feel like I've just been shown an example of how I can go about doing that. I think I just learned a way to increase the overall value of the community or region while leveraging what is available in the community.

Furuta: In my eyes, Mr. Kamiyama is a lobbyist—an unusual one, perhaps—and he himself is also a business player. Not to mention that he has put aside his own business to communicate to others about this while keeping a level-headed perspective about the overall industry. I think he is probably aware that if he moves in a certain way, the people engaged in related businesses would have no choice but to move in that direction as well. And this is probably why he's doing all these things. But even so, I would think that his Osaka dialect must have somehow helped him to push through this complicated and bizarre world of conflicting interests. The pivotal part is the Osaka dialect! [Laughs]

Kado: I teared up a little as I listened to Mr. Kamiyama's story because there were many parts that overlapped with my own. When I was driven by a sense of mission and started doing these things, I got so wrapped up it that I sometimes wondered what exactly I was doing.

Takemoto: I think there should be one Mr. Kamiyama in each industry. [Laughs] This kind of constructive and productive way of moving things forward is something they don't teach you in school. When I watched the interview footage, the words that struck me the most were "I'll get it to collapse from the inside." He has made clever use of his Osaka dialect, which is ambiguous in its level of formality, to create aggressive industry associations while at the same time getting those with vested interests to take his side. I was impressed by the fact that he used all possible means to accomplish his goals.

Murakami: He was, in a sense, in a position to control the system, except that his goal was to come up with a plan to get it to collapse and rebuild the system from within. And I think that is wonderful.

The importance of the human network

Hori: The former national government officials that we talked to also said that now is not the time to act in the interests of ministries, and indicated that they would like to work with us in various aspects. What do you make of these changes in these people who are at the receiving end of lobbying?

Kamiyama: National government officials are often treated as bad people or seen as a privileged class, but I don't think that's the case at all. Many of them are thinking about the future, have the same level of sensitivity or perception as ordinary people, and are of course proficient in discussion. That was why I thought a good way to get my foot in the door would be to talk with government officials and organize the key points of each case. This makes it easier for us to discuss, and properly lining up the facts also makes it easier to talk to National Diet members as well.

Kado: I have many opportunities to talk with administrative officials as well. And even though I can talk about the specifics with each and every one of them, I sometimes still find it extremely difficult to draw an inclusive conclusion at the end. This is especially the case when a project involves multiple ministries; even if the idea is good, it stops at the question of, "Who's gonna do it, then?"

Murakami: It is true that administrative officials of the national government are good at organizing an issue and generalizing it to an abstract concept or idea, and then coming up with steps to give it shape. But it is difficult for them to visit the site in question and start making arrangements for action. Administrative officials, in particular, cannot be really expected to respond or move quickly enough to deal with issues that they cannot handle or resolve within their own organizations. To physically visit the site in question and come up with specific actions beyond the organizational level, even administrative officials need a high level of expertise.

That being the case, it will then become crucial to find people in the private sector who can execute the drawn-up arrangements and then work together immediately to roll out the rest of it. The most pressing issue now is not only what to do, but also who will get involved in the actual work. A network of people with relationships you can trust that transcends public and private sectors is extremely important. And that is why we are discussing this now.

Kamiyama: Yes, that's exactly right. Both the person who pitches the ball and the person who catches it and runs are important. That is why national government officials are equally important, and I myself also agree with them that we have to "give it shape." And when it comes to making up rules, it is the politicians who get to make the final call, after all is said and done. If this is the case, then someone has to lay the groundwork, set the stage, and make all the necessary preparations in advance, as well as coordinate opinions and iron out the differences.

Even when talking to industry associations, I have no intention of fighting with them or confronting them just because the regulations are relaxed. We have become comrades and get along well with all the heavyweights.

Some of you may think, who can do this, then? Well, anyone can do it! But there is one thing that you must absolutely do at any meeting. And that is to arrive before anyone else, take up a spot or seat right at the front, and pay attention to what everyone has to say, in a way that everyone can see you. And then you go around and politely greet every single person. This is important. If you want to achieve your goal, you have to do all that, no matter what. You should also keep accurate notes of who you talked to and what you talked about. Even though it may look unremarkable, if you continue to keep at it, day in and day out, people will start to remember you. Unless they remember you, you will never get the opportunity to meet or talk to them. And as you steadily build relationships this way, you will gradually be given opportunities to speak at official meetings and gatherings. It is the accumulation of all this legwork that will bring you to this point.

Linking policymaking to implementation

Hori: Earlier, Ms. Fujisawa asked if anyone other than Mr. Kamiyama could do what he does. How do all of you feel after listening to the discussion so far?

Fujisawa: About 10 years ago, I had the idea that I needed to get myself involved in policy if I wanted to change things. So I talked to experts and tried to take action on my own. But in the end, I thought that it was a bit too hard for me.

If you don't carefully and thoroughly communicate with each and every person, you might upset the balance in terms of who is on whose side. And when that happens, you won't be able to get anything done anymore. But if you try too hard, you may come across as someone who has ulterior motives and consequently lacks impartiality, and the gears of the project or initiative may stop moving. Yet Mr. Kamiyama has been able to move with precision and finesse in this delicate and complex environment. The extraordinary strength of his conviction is unbe-

lievable, and I have nothing but respect for him.
Maki: I don't have the fortitude or character to do that kind of thing either, but it is very reassuring to know that there are people like Mr. Kamiyama out there. There are many different players engaged in rural revitalization; a lot of them truly want the best for the community, and there are also those who have their own fortes and abilities. And in my opinion, it is extremely important that there be a way for us to find and approach these people.
Kado: I think this is exactly what it means to move back and forth between the specifics and the abstract. There are ideals and plans, but there is also reality and the actual situation on the ground. Unless we produce more and more people who understand this and have the power to implement it, I don't think rural revitalization will ever be possible.
Abe: I feel that there are roughly three elements to get policymaking and implementation to work, after listening to various people and sorting through the case studies I've collected.

The first is to gather idealistic people who have expertise in administrative and clerical work. Such administrative and clerical workers often gradually lose their idealism or beliefs and become mere robot-like workers. However, I am of the firm belief that it is important to gather as many idealistic staff as possible that will neither lose sight of their ideals or dreams nor become mere "robots" in your camp.

The second is to ensure that when the government and the public administration are willing to do something, they will not receive an unenthusiastic response. That is what the policymakers are most afraid of. To prevent this from happening, we must encourage players who will implement the policy to proactively express their interest in the implementation. I'm actually talking about the private sector here. We need to make sure that this is done.

The third is to have a certain amount of power to be able to rebuke, persuade, or placate those who disrupt the conversation at the end for some reason.

Everybody will have their own way of doing things, but as long as these three elements are in place, I would say that about 50 or 60 percent of the work can be accomplished. Mr. Kamiyama is amazing in that he encompasses all three elements.

The key to moving local governments: legitimacy and legality

Hori: How do you deal with the heads of the local governments?

Abe: When you commit to a community or region, you have no other option but to align yourself with a faction. For the initiative or project to go ahead, it is necessary to create a system that will allow the head of the local government to continue with it. One way is to provide indirect support from the outside, and the other way is to structure it such that the initiatives can continue on their its own even if changes are made to the system. These are the two types of systems that I can think of.

Takemoto: Yes, I think it is necessary to more or less create a system, since the head of the local government is bound to be replaced someday. But for us to get the head to understand and give the green light for the project, or at the very least not get the red light, we cannot avoid dealing with the head in person. And it takes a tremendous amount of effort and stamina to build those relationships.

And that is why I believe that legitimacy and legality are critical. First is legitimacy in terms of the position we take in dealing with people and projects. After we have secured that legitimacy, we need to then explicitly demonstrate our legality through reliable information or new policy process(es) so as to create a situation that will better allow us to convince and win over those who are confronted with the situation.

Furuta: For myself, I tend to take the approach of creating results first. For example, when I launched the "Morning University of Marunouchi" for working adults, I was told that it would not attract anyone. But since I didn't have the time to persuade people, I started off by paying for it myself. It eventually took off when I disclosed our achievements, and we attracted thousands of people in the end. This is the same for crowdfunding, too: as soon as you reveal how much money has been raised, people's views start to change.

Kamiyama: There really are various ways to go about doing it. I also showed them the opinions of ordinary people in the form of quantitative data and told them we should take action because that's what the market data said. But in the meanwhile, I also made use of the national government to convince the community that we had to do this now, citing reasons like the national policy is such and such, or that the ministries want us to do this and that.

In any case, we must make sure that the head of the local

government is on the same page with us. But this doesn't mean that everything will go without a hitch just because the head has given us the nod, does it? Life is not that smooth. It is important to talk to all kinds of people, including those in public offices, those working on the ground, and even those who speak out against it.

Giving love to every player; deep convictions

Hori: So how can we come out top, then? Let's hear from Mr. Kamiyama.

Kamiyama: I think it's best not to go head-to-head with the opposing forces.

Abe: For regular discussions or debates, the one who is right to some extent will win. So it is always better to work on it in an open forum. Mr. Kamiyama has a sound argument, so when push comes to shove, he can say, "Let's do this openly." And I think that is why he has been able to push through and ahead with it. In other words, what is important is an open discussion. This is also an important role and function of the media.

Kado: The opposing forces whose arguments do not hold water work behind the scenes, never in full view of others. The more we push ahead, the more they back away.

Takemoto: In my case, and probably Mr. Kamiyama's as well, we are working with a sense of mission and conviction that this is what we should do, so it has never been a matter of fighting or not fighting. I feel that I can always speak my mind to the people that are facing me because I have openly declared my noble cause, so as to speak.

Kamiyama: And that is because we are convinced that we are right.

Takemoto: Yes, absolutely. [Laughs] It's not a win-lose approach, so I don't waver at all.

Kamiyama: For example, when I was asked to attend a meeting of the Liberal Democratic Party to discuss the Private Lodging Business Act, there were people who said all kinds of things, but I chose not to defend myself. I just kept saying the same things over and over. I stayed cool and composed and let them think that I couldn't read between the lines. I don't know if that was a good thing or a bad one, but I think it turned out to be good if we look at the results.

Furuta: Although Mr. Kamiyama may look like he is rattling on and on in Kansai dialect, he is actually speaking clearly in a well-thought-out

manner. The reference materials on private lodging and *albergo diffuso* he created are well organized, so they are easy for administrative officials to remember. I'd say that one of the keys to his success lies in his meticulous preparation.
Takemoto: That's true. The fact that he doesn't talk like he's an eloquent speaker also makes him come across as a trustworthy person. Although it may look like he is very inept at speaking, I'm sure he has it all planned out in his mind. [Laughs]
Miyase: He just appears on the scene in a really sketchy way, but when he starts talking and drawing you into the conversation, it makes you think, "Oh, this guy is serious."
Maki: And Mr. Kamiyama is deeply connected and working together with the people in public administration as well as those busy on the ground in communities, and they are moving forward with their eye on the future. I find that very moving.
Fujisawa: I think Mr. Kamiyama is always thinking about all the things that need to be resolved and working on it so that it's easier for everyone involved to do their jobs. That goes for the mayor, national government officials, and those on the ground. He is trying to figure out how to support them to make it easier for them to drive the conversations. As he paints the big picture, he also looks at each and every person on a microscopic level and pays great attention to them. He has a delicate sensitivity and deep convictions. And I have just realized today how wonderful these two things are.
Murakami: On one hand, he is very firm about his beliefs, and on the other hand, he has the ability to be open-minded and listen to the assessments and evaluations of the community that supports his beliefs. One could say that he is a person that has both qualities.
Miyase: During the interview, Mr. Kamiyama said something that left a deep impression on me. "I do all these things with love for this community. But since there are also people who love this community more than I do, I want to create an environment where they can express their love and affection. That is my job, and that's what makes it worthwhile."

Session 4: Ms. Kumi Fujisawa's Case —Kamishihoro, Hokkaido

A shift from a distribution mindset to an investment mindset: Disclosure, governance, and finance

"Rural revitalization means investment. Yet in rural areas, the concept of investment is extremely poorly understood," argues Kumi Fujisawa (Institute for International Socio-Economic Studies). From the viewpoint of a finance expert with a proven track record in the world of investment, regions that have succeeded with rural revitalization all excel in the three fundamental elements of investment: (1) disclosure (information disclosure and dissemination); (2) governance (project responsibility management); and (3) finance (fundraising). This discussion will be based on the case of the town of Kamishihoro in Hokkaido, which has successfully increased its population by attracting quality talent and companies through a variety of projects.

How and with whose money can businesses be created?

Hori: 'Today's themes are disclosure (information disclosure and dissemination), governance (project responsibility management), and finance (fundraising). The discussion will highlight how these three elements, which are common sense in the investment world, are also the three pillars supporting rural revitalization. The presenter is Kumi Fujisawa. Her experience encompasses founding Japan's first investment trust evaluation company and anchoring the NHK Education TV program "21st Century Business School," in which she traveled around Japan conducting interviews.

Fujisawa: Through those interviews, I was able to discover that the future of Japan will be transformed by people who are serious about working in rural areas. I learned that rural areas are interesting, and even today, as a board member of Shizuoka Bank, I continue to support rural areas. I am also helping the city of Kawasaki publish a book about its initiatives as a means of popularizing them across Japan.

Mr. Murakami posed an amusing question for me today: "What is a

Fig. 1-5. Why does some rural revitalization advance while some stop?

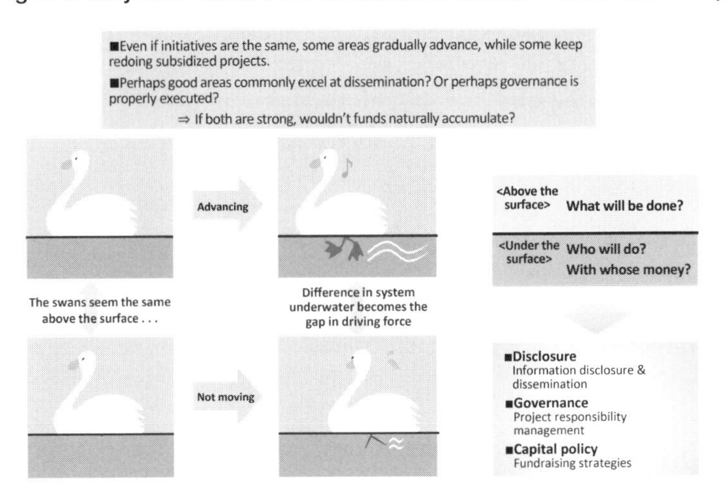

swan doing under the water?" The swan is a metaphor for rural revitalization projects. Most swans are elegant, beautiful, and possess a noble purpose. Yet in reality, they often hardly move forward. And when we peer into the water, we discover that while they appear magnificent above the surface, they are only paddling gently below it. On the rare occasion when we see a swan moving forward with vigor, we will also find that it is paddling energetically under the water. Where does this difference in the paddling under the surface of the water come from? That's what you were asking me, right, Mr. Murakami?
Murakami: Yes. The hypothesis is that the underlying causes of the difference reside in the areas of disclosure, governance, and finance.

Investors, while certainly concerned with the substance of a project, place even greater emphasis on who is administering it, under what structure, and with whose money when deciding whether or not to invest in it. The key to whether a regional project succeeds or fails may actually lie here as well. It is also important to make sure that information is properly disseminated. If you work hard to communicate what you want to do or are doing before you get absorbed in thinking about what kind of project to pursue, you will attract money effortlessly.

When we look closely at a swan that isn't moving forward much, we find that even though the project was launched with the help of a subsidy, it's proving difficult to attract the human resources needed to implement it. Or the subsidy has all been spent, making it impossible to translate the outcomes into a commercially viable business. So what is a swan that's moving forward splendidly doing underwater? How has it strengthened its legs and feet? In this installment, Ms. Fujisawa, an investment professional, gives us her assessment.

Three peculiarities that are common to regions

Fujisawa: The example I'm going to share with you today comes from the town of Kamishihoro in Hokkaido, but before I discuss it, I would like to talk about three peculiarities that are common to regions that I have noticed in the course of working on the various projects I have been involved in.

The first is carelessness with money. Getting hold of money is very difficult, isn't it? I have experience starting my own business, so I know how tough it is to launch your own company and get people to buy what you have created. But when I go to the regions, I meet people who complain about having no money, yet I get the feeling that they don't understand the pain of actually getting money. Because they haven't experienced the sense of gratitude that comes from receiving money, they are sloppy in how they use it. They don't look like they are using it carefully. And I've had the impression that rather than respecting people who are making money, they tend to feel envious and jealous of them.

The second is failure to put money to work. Their money could even be described as "dead." The money just disappears as though it were being absorbed by a sponge. Ideally, a mechanism by which money generates more money should be in place. For that to happen, calculations that take account of anticipated risks would need to be performed, but no such calculations are being made. They don't even consider what sorts of risks exist. All they have is a vague sense of crisis, saying things like, "Our region's in big trouble." It's as though, rather than thinking about what they want to do in the future, they're just focused on today: "Things are terrible right now."

The third peculiarity is that they don't trust people from outside their own region. Having said that, if someone comes up from Tokyo, they trust them. Even more so if it's a famous person. When that happens, they're in fantasyland. They also have the illusion that someone

who gives them money is a good person, and someone who's more careful with their money is a bad person. I don't see any signs that they're trying to think about the meaning of money. Furthermore, the locals tend to regard volunteers as free labor and exploit them. The locals may never think about why volunteers have come or what they should learn and take back with them. And finally, they end up with the misapprehension, "Outsiders wouldn't understand even if we told them, so why bother?"
Hori: Mr. Kado, you kept nodding as you were listening.
Kado: Yes, because I totally agree. And I would even add something. If all that were reversed, the regions would start moving in the right direction. The question is how to manage that.

Gather money after thinking about what to use it for

Hori: Now I'd like to tell you about Kamishihoro. Located in the northern part of the Tokachi region of Hokkaido, it's a compact town with a population of about 5,000. It's blessed with a rich natural setting, as it lies at the foot of Daisetsuzan National Park, and more than 70 percent of the town is covered with forest. Agriculture, dairy farming, and forestry are thriving; the town is also famous for the Nukabira spa resort, where the baths are filled with natural hot-spring water.
Fujisawa: The hot springs there are truly wonderful. The water makes my skin super smooth. The location is also incredible. The stars look like they're about to fall from the sky.
Hori: While enjoying such beautiful scenery, you take an almost perfectly straight farm road from Tokachi Obihiro Airport to the center of the town. There, a demonstration experiment is now underway to operate Japan's first self-driving bus on a public road.
Fujisawa: When I first visited there, I was told, "When you exit the airport, go straight down the farm road and turn left at the first traffic light, and you will be in Kamishihoro." But in fact, that first traffic light is a 50-minute drive. [Laughs] It is a very peaceful place.
Hori: Kamishihoro has been suffering from population decline for a long time. What turned things around was the Hometown Tax Program. Instead of spending the tax revenues each fiscal year, the town saved them up in something called the "Kamishihoro Hometown Tax Dream Fund for Childrearing and Tackling the Declining Birthrate." It then

The Taushubetsu River Bridge is one of the famous tourist spots in Kamishihoro.

started systematically investing money from this fund in children's education. This was apparently the first such fund in Japan, and it halted the decline in population—the population is actually increasing now.
Fujisawa: That's right. I started working with the town just after they had started their Hometown Tax Program. The population had just begun to edge up. What I thought was great was that before they launched their Hometown Tax Program, they decided exactly what they were going to spend any money they raised on. In most cases, the money raised tends to get distributed to those with vested interests. But they were aware of that, so the town council decided in advance to use it for education, as there was unlikely to be much opposition to that.

The first thing they did with the cash was to make certified childcare facilities free of charge. Then they began providing high-level education at these facilities by bringing in expert teachers in the fields of art, English, and PE. They then did the same with elementary schools, and as a result, the children's academic performance climbed. I hear it's now above the Hokkaido average.

Subsequently, certified childcare facilities became free of charge nationwide thanks to a change in the law. So now the money that they had been spending on the childcare facilities was freed up, and they were able to allocate it to other strategies. So they're always looking two steps ahead when thinking about how to use their cash. This is something that is worth highlighting about the mayor and other individuals in charge of Kamishihoro.

Wanting to change the town through education and achieve substantial growth

Hori: The mayor is Mitsugi Takenaka. He's currently serving his sixth consecutive term. He takes on various new challenges with a sense of speed, doing everything himself, doesn't he?

Fujisawa: He certainly does. Even when he was talking about the increase in population, he used PowerPoint materials that he had produced himself. He always makes what's in his mind visible as he explains it to the people around him. That's how he develops his concept for the region, through this process of producing and talking.

My connection with Kamishihoro started when I interviewed Mayor Takenaka for my work. We met when Kamishihoro held its first event in Tokyo for people who had made donations under the Hometown Tax Program. Many people, including families, came to that event and enjoyed the free food and drinks that they received as a return gift. Among the offerings was Tokachi Naitai Wagyu (Japanese beef), which is extremely delicious. This tasty beef was provided in exchange for coupons, and everyone was lining up to buy the coupons. So even though it was ostensibly a thank-you party, they were even managing to raise money there.

What I thought was really clever was the migration consultation area located on the way out. Having been shown how wonderful Kamishihoro is with various videos, presentations, and goods, the visitors found themselves in this area just as they were about to head out. For some reason, everyone sat down there as through they'd been sucked in. They were told that they could experience living in Kamishihoro on a trial basis, as there were houses where they could stay for free. It felt like an invitation to stay for a few days as an extension of a Hokkaido trip.

Hori: Mayor Takenaka was originally the education chief, right? I've heard his main purview was lifelong learning. Since he came from the field of education, not finance or agriculture, people around him viewed him with skepticism when he became mayor. He would apparently

often be asked, "Do you understand economics? You don't know anything about agriculture, do you?" He said that each time this happened, he would explain that he was going to "grow the town through education."
Fujisawa: That's why academic performance improved so much. However, the town doesn't have a high school. Children have to go outside the region for high school, so Mayor Takenaka says he wants to do something about this next. And since the money was initially spent mainly on children's education, there were also requests from the elderly, so after a while he opened something called the Kamishihoro Lifelong Learning Center. Declaring, "Elderly people will also be happy in this town," the mayor took the lead in this initiative to provide opportunities for learning and engaging in activities so that elderly people can have a healthy, happy life. Whenever I visit Kamishihoro, the center serves as my base for interacting with the people of the town.

As this initiative shows, the mayor is truly one of those rare individuals who always comes up with a grand vision, devises a process to realize it, and continues to produce results one step at a time.

Becoming a place that attracts more and more young talent

Hori: Ms. Miyase and I went and interviewed Mr. Takenaka about what his approach to town administration has achieved. Kamishihoro is a really small town, but there were all sorts of things there.

For starters, there's Michi-no-eki Kamishihoro, a roadside station that just opened in June 2020. The fashionable atmosphere is impressive, with cafes and restaurants lined up alongside agricultural-product sales outlets. What is really noteworthy is the company that runs it, Karch Co., Ltd. It's a regional tourism trading company that's a 50-50 public-private partnership.
Miyase: Yes, in addition to the roadside station, it is also pursuing various other businesses, setting up a destination marketing organization and promoting biogas power generation, for example. Being a dairy-farming town, the company is also promoting local production for local consumption of energy by using cow manure, which is a waste product, to produce electricity.
Hori: Such diversification in business has proven successful, and it's been in the black since it started operations. The company receives cash in the form of "designated administration fees" from the town to help

with the upkeep of the roadside station. The plan is to gradually reduce these payments such that by the fifth year the station will be able to stand alone as a profitable concern. So it continues to resist calls from local people to give them bigger price discounts, because its mission is to become financially independent. It explains this proudly to the local people, and as a private business, it has pledged to aim for a sustainable bottom line while making business decisions in a speedy fashion.
Miyase: One of the company's division managers, who originally hailed from the tourism association, said that this would be a way to give back to the townspeople, who are effectively also shareholders.
Hori: There is the Kamishihoro Share Office, which also opened in 2020. This is a co-working space created for people who visit the town for work; its tagline is "A new way of working: Sharing the city and the countryside." The people entrusted with the launch and operation of the space were young people from all over Japan who applied to join the Local Vitalization Cooperators, a program endorsed by the Ministry of Internal Affairs and Communications. When I actually visited the place, I found those young people confidently experimenting with various new things.

For example, they are providing elderly people with information and support services via tablet PCs. A demonstration experiment was conducted in Kamishihoro that involved every elderly household being given a tablet free of charge and taught how to use it to reserve welfare buses, obtain meal services, and manage their health. To enhance usability for the elderly, the screen design was improved by making the buttons on the display larger and eliminating the need for swiping, making it as simple to operate as a touch panel. And it was one of the Local Vitalization Cooperators at the co-working space who made these enhancements.
Miyase: The Local Vitalization Cooperators are also helping people launch businesses and offering advice on places to work, so I came away with a strong impression that this is really a place where talent comes together.
Hori: Another recent example worth paying attention to is a smart-logistics initiative involving drones. Hokkaido, of course, is vast. So the cooperators decided that to transport and share various items, a hybrid approach involving both land and air should be employed. This led, in

August 2021, to Kamishihoro and three companies (Seino Holdings, Dentsu, and Aeronext) concluding a comprehensive cooperation agreement for "sustainable future town development" utilizing next-generation advanced technology that includes drones. A demonstration experiment where food, specimens for diagnosing disease in cattle, cow embryos, and other items are transported by drone has since begun. Transporting cow embryos by air is a world first, isn't it?
Miyase: The idea was to use a drone to transport the embryo to the cow, rather than moving the bull to the location of the cow to be inseminated, which takes time and causes stress to the bull.
Hori: The aim is to reduce stress and increase the probability of birth while also reducing the environmental cost of transporting the bull in a truck. More and more experiments like this are getting underway, and because they are being conducted by younger people, start-up entrepreneurs, and employees of large urban companies, more and more people are coming and going in the town. I could sense a very vibrant atmosphere.
Fujisawa: My thoughts exactly. Mayor Takenaka also welcomes them with a "come anytime and do whatever you want" vibe.

Why does top-flight talent gather in Kamishihoro?

Fujisawa: A lot of people are applying for the Local Vitalization Cooperators, right? And my impression is that from among them, Kamishihoro has selected the very best and brightest. I am very interested in how they find such people.

Besides, in this town, they say, "You can do anything you want," but at the same time they also say, "We won't give you money." This means coming to Kamishihoro and trying something with your own money. While this may be appealing, it makes for a substantially higher hurdle. How do they connect with people who have that kind of spirit and energy? This is also a simple question.
Hori: The purpose of the interview was to find the answers to Ms. Fujisawa's questions.
Miyase: In addition to Mayor Takenaka, one of the key people who knows the answers is Toru Kaji of the Digital Promotion Section in the Kamishihoro town office. He has been working with the mayor for 20 years, and as his right-hand man, has been involved in the Hometown

Tax Program and efforts to promote the town using information and communication technology (ICT). He was selected to head up the ICT Promotion Office this past April.

Fujisawa: He's someone who can see the future clearly, get to the point, and communicate in his own words. He's also someone who has a proper understanding of money.

Miyase: The Local Vitalization Cooperator I mentioned earlier—the one who helps administer the co-working space and developed the interface for the tablets—was also hired by Mr. Kaji using the scout function of a job site called Wantedly. It is unusual for a local government to allocate money for a scouting drive that may or may not be successful. However, Kamishihoro funded the scouting effort by Mr. Kaji, who also felt it was a gamble, but the person eventually hired through that effort said of Mr. Kaji: "He's a very enthusiastic person, and I really wanted to work with him."

This individual, who had been living in Tokyo but would now be returning to their hometown, reflected that meeting Mr. Kaji was a big reason for that. "It made me realize that with this type of work, I could maintain a connection with Tokyo even while living in Kamishihoro. . . . It gave me an experience of turning something from 0 to 1." So the town is able to attract active young people by showing its willingness to take responsibility for bringing them in.

The soundnesss that comes from the "we won't give you money" position

Hori: Kamishihoro has focused on using various techniques to increase the number of connected minds. The same can be said for its efforts to attract companies through things like demonstration experiments.

Miyase: That's right. An example of that is Nippo House Kamishihoro, a workcation facility that opened in April 2022. It was built with public money, but is privately operated. The facility, which provides accommodation for employees dispatched to the region by companies for medium- to long-term stays, was created with support from Spicebox, a Tokyo ad agency, and Muji House. According to Dai Yoshida of Spicebox, who oversaw the design and handles the operation of the facility, he became acquainted with Mr. Kaji through work related to the Hometown Tax Program. Mr. Yoshida mentioned that his company wanted to do something with a municipality that was proactive about sustainability as part of a branding initiative for Muji House, to which Mr. Kaji replied, "Please make it Kamishihoro."

Spicebox is contracted by the town to run Nippo House as its designated administrator, but does not receive any administrative fees from the town—in fact, quite the opposite, as it pays rent to the town. This is in line with what Ms. Fujisawa said: The town lets them use the space, but doesn't give them any money. That being the case, why do companies go to Kamishihoro? Mr. Yoshida explained the reason for that in incredibly clear terms.

> The normal approach would be to receive designated administrative fees for a while, and during that time, ascertain whether it will be possible to make a profit. Given the risks involved, that makes sense. However, as an advertising agency, we felt that perhaps we shouldn't be limiting our activities to the Tokyo metropolitan area. We were concerned that we were looking only at people in Tokyo, even though our advertisements are targeted at the entire country. To rectify this, I thought it was important for us to go out to outside regions and work and interact with people there so as to tap into their local perspectives.
>
> However, we would require the cooperation of the local government, and we didn't want the local government to become overly dependent on us. We wanted the company and the local government to both take on various new challenges, almost as though we are competing with each other. Kamishihoro was immensely attractive as a place where we could build such a relationship.

Another reason he was drawn to Kamishihoro, Mr. Yoshida says, was "because it's fair." With most local projects, the execution structure is designed to reflect various "grown-up" factors, such as relationships with related businesses. Yet in Kamishihoro, rather than just appointing Mr. Yoshida and his team, who had already established a relationship with the town, they publicly invited bids to decide who would build this publicly constructed, privately operated facility.

> Despite our close relationship with them, having given them lots of advice and having had lots of talks with them about the idea to build such a facility, there was still a possibility that our proposal would not be adopted. I was surprised by their commitment to fairness.
>
> Moreover, I could perceive this mindset in the townspeople and in the council members. So the one-year preparation period involved some really hard work, but when that work finally came to fruition, everyone was delighted for us, which was very moving. There was no gray zone at all, and that was great.

Mr. Yoshida says that Mayor Takenaka is the prime example of this. Apparently, he drops in to Nippo House every morning and asks questions like "Are you getting customers?" or "Do you have any problems?" He did that just after the facility opened, and continues to do so every single morning. The mayor told him he takes a walk every morning for exercise, and uses that as an opportunity to pop in to all the facilities in the town and see how everyone's doing.

What has been Mayor Takenaka's motivation for getting involved in town administration? Here are his own words:

> One of the most important things for rural revitalization is the disclosure of KPIs [key performance indicators]. This is something that has rarely been done by governments, but it must be done. The reason they don't want to do it is because they are afraid of the consequences. But if something goes wrong, they are going to have to disclose it, so they should be open from the beginning. In Kamishihoro, everything is open, even in council and committee meetings.

Money generates more money, and people attract more people

Hori: A person from one of the companies told me that the mayor and Mr. Kaji would swiftly reject companies that they could tell were coming to them for grant money. I got the impression that only companies that make a serious presentation get to stick around. The town's secret to "never running out of money" also seems to lie here.

Miyase: I would agree. Mayor Takenaka really hates the idea of just throwing up a building and being done with it. So he tours the town every day to check that everything is running smoothly. His stance is that even if an initial investment is funded by the national government or the town, it is the recipient who is ultimately responsible for making the investment work. I found it be a town where tension is channeled for good use.

Fujisawa: The town has become a place where companies can verify the viability of their business ideas. I get the impression that the companies are looking to eventually expand their businesses nationwide. In other words, they are "serious" people. That's why the town entrusts them with its projects and provides them with a venue for taking risks. And the key point is that the money invested is not dead, either.

What's more, not only does money bring in more money, it also brings in people. They don't come to hang out, but to challenge themselves. I felt that this was a place where money thrives and people thrive.

Maki: The young people at the town office look great, don't they? Regional initiatives often seem to be hidden below the surface, but they struck me as being very open and above board.

Takemoto: To what extent should townspeople be informed and enlightened about what's actually going on? In other words, I thought they were practicing information disclosure in its most fundamental sense. In the case of the town of Ama, which we talked about last time, the boundary between the public and private sectors is more ambiguous, and there's a visible commitment to working together. The population is also small, at just over 2,000, so it's impossible to do anything nefarious because everyone can see who is doing what. The two towns each have their own way of disclosing information, but I think there are similarities between them.

Kamiyama: I think this is the only direction for municipalities with small populations that are thinking about what to do going forward. There are many cases where a local government has constructed a building, but the project is still in the red and not doing well. I think that if it were managed properly, it would probably be viable.

Furuta: I hold the same view. I think one of the reasons things don't go well is that the people who construct the buildings and the people who operate the businesses inside them are being deliberately separated. The local government may be trying to keep them separate to maintain fairness and to avoid total control by a single company, but I think they should take the opposite approach if they want their projects to succeed. Unless they delegate responsibility for everything to the same company from start to finish, they won't get results. But they don't want to take responsibility for that. In the case of Nippo House, we were able to get a lot done because ultimately everything was left up to us. With regional projects, I think it's important to delegate full authority.

Hori: Discussions by the town council also got pretty heated in the stages leading up to that. There were times when council members blamed a town official, saying things like, "If we start this kind of project, it will cost a lot of money to run and we won't be able to keep things under control." But the town official replied, "No, you're wrong. We only make the initial investment. After that, the private-sector people will make the business sustainable. We absolutely never deal with people who are just looking for money." I was impressed by this clear stance.

An oasis free of "encumbrances," in which everyone can operate on a level playing field

Abe: I felt this was the "royal road of royal roads" for rural revitalization. I believe there is a royal road to victory in rural revitalization. It is to create a situation in which the rural side—namely the local government—can choose its partners. It's also limited to pioneering cases. There have been very few subsequent cases where the local government has been able to select its partners. A pioneering case obviously has no precedent, so you have to think hard and be prepared to respond to any questions or criticisms that may arise. I think this leads to the seriousness that Ms. Fujisawa is talking about. Kamishihoro is executing this royal road strategy very shrewdly.

Its utilization of the Hometown Tax Program is smart as well. Hometown Tax donations capitalize on a very small portion of the user's disposable time. A moment's effort from the user causes money to flow into the region. Kamishihoro leverages these touchpoints by inviting contributors to thank-you events that require a bit more of their disposable time, which in turn provide an impetus for sightseeing trips, which swallow even more of their disposable time and can even lead to them moving to the town. The components of this strategy are seamlessly connected, creating gradually higher levels of engagement with the town. It's a winning pattern for rural revitalization.

However, initiatives by local governments are often siloed, and lack a consistent approach. I think this is why they fail to succeed.

I also think it's great that they're utilizing a fund that allows budgets to be spread over multiple fiscal years. Scouting talent is another good approach. Since responsibility for the results of using money is clear, there's no vagueness like what is typically seen at government offices. Furthermore, the mayor himself checks on progress with the projects: "How are things going with this now?" He's always monitoring progress, which is fundamental to management. Of course, the size of the town's population is ideal for that, yet while that might be a factor, my impression is that he's doing everything he possibly can.

Murakami: People from companies that have come to the town have said that they did so because it's fair, and those comments have really resonated with me. It's a region that has successfully made itself an oasis free of "encumbrances" (i.e., nepotism). There's a feeling of openness, and no sense of confinement. They are resolutely on guard to prevent any hindrance by "encumbrances." They say, "We won't give you money, but we'll treat you all the same." Everyone comes here in search of that sort of integrity.

Ultimately, it's impossible for a municipality to do everything 100 percent on its own. The presence of partners is absolutely essential. These partners bring new value and bear a portion of the costs. I thought it was fantastic that they have created circumstances in which this is possible and also ensured a fair environment.

Rural revitalization means investment

Fujisawa: I thought that a common theme running through everyone's stories is whether one is "serious" or not; in other words, where responsibility resides. Kamishihoro has made clear that it will take responsibility. And they take things forward only after having first predicted the future and designed the entire project. I think that's why things are going so well for them.

My general feeling about regions is that things taken for granted in the investment world don't exist in rural areas. I believe that rural revitalization is actually an act of investment itself. You're starting something new, so it inevitably involves taking on risk and investing capital for growth and returns. In other words, it's investment. The investment is not limited to money; it also includes people and things.

Investment is not consumption. It is for the future. There is the term "ethical consumption," and this can be viewed as a type of investment, as it is aimed at changing society. Even lifelong education is an investment, as it involves developing oneself throughout one's life. The same goes for the circular economy and ESG (environmental, social, and governance factors). The investment mindset is essential for a sustainable society. However, neither an investment mindset nor investment behavior can be found in most rural locales.

In the investment world, it is generally accepted that the first step is to define the investment project and assemble the investment team. This process clarifies where responsibility will reside. The project is designed to reflect where there are risks, and the magnitude of those risks, and how they can be minimized. Naturally, there is the issue of responsibility structure, and it's necessary to consider how to secure returns while taking on an acceptable level of risk.

The next step is to decide how to raise money and from what sort of people. In other words, it's to choose the investors. It's no good to just ask for money without caring about who's supplying it, because the required return and tolerable risk level differ depending on the person. So you only select investors with the same goals in terms of risk and return as yourself.

The final step is to accurately report, down to the last fraction of a cent, how the money entrusted to you has been used and what the results were. By doing this periodically, you prevent investors from pulling their money out. It also helps you to attract more investment. That's the investment cycle.

If you don't follow these steps, the money will just gradually disappear. Yet that's the reality in rural areas, I feel. With one regional project, we worked hard and finally got a grant, but when we asked local people for ideas about how to use the money and about project strategies, we were told, "We don't care about strategies, as there's no knowing whether they will be successful, anyway, so just distribute the money to us right now. We'll make sure to spend it for you." I was shocked when I heard that. In the end, we didn't just "distribute" the money, but this episode still goes to show that the bulk of subsidies is consumed, not invested.

Going back to today's theme, what this means is that the governance and disclosure needed for starting a new business is absent. Governance is about clearly assigning responsibility. Making it clear which people are serious—or, put another way, which people are going to be taking responsibility—is the most critical component of governance.

A mechanism for getting investors to participate in a second round

With disclosure, meanwhile, the most important factors are accountability and transparency. It's about more than just reporting results. Interim reporting is vital, and even before the money is raised, you need to state what you're going to be using it for and what sorts of outcomes you're planning to deliver. These factors are expressed numerically, but I also believe there needs to be a story that is going to move potential investors to put their money in.

If you take the attitude that using someone's money means you are responsible, all this comes naturally. However, when they're getting money from the national government or a local government, people don't do these things. From the perspective of investment, it is really irrational.

Hori: In Kamishihoro, they're following those principles to a tee, aren't they? They send letters to people who donated money via the Hometown Tax Program, and also try to provide lots of information on their website. And when hiring the Local Vitalization Cooperators, apparently it was the mayor himself who initially issued the order to get people who are skilled in information dissemination.

Fujisawa: Indeed. Communicate thoroughly at all stages: before, during, and after. Doing so actually makes people want to invest again. Creating a mechanism through communication that makes people want to get involved again leads to a sustainable increase in funding and a continuous rise in the number of connected minds. Kamishihoro seems to have grasped this key point. After all, the mayor himself has been taking the lead in doing this. He produces material himself and constantly updates it, so every time you meet him, there's something new. So everyone can envision the future he's trying to create and share his dream, I think.

Abe: Exactly. As you say, the first thing a start-up firm should spend money on is public relations. First, increase the number of connected minds, then gain recognition in various circles, and finally, use that as leverage to launch the business. And I think the same applies not only to startups, but also to rural revitalization efforts.

Fujisawa: "Investment mindset" might sound complicated, but as Mr. Abe points out, whether it's rural revitalization or starting a business, the principle is the same for any project. There are two types of people in the world: those who want money and those who want to invest money, but there are sometimes governments that provide and use money in peculiar ways. It's important to skillfully create relationships among the three parties, and this should also leverage the power of public relations.

Ways of getting money include accepting donations, taking out loans, and receiving investments. And within the investment category alone, there are various types. There are also myriad ways of returning value to the people who provided the money. Some people want accolades, some want steady interest income, and some want to make money themselves as participants in the project. Some people just want to be involved in the project, even if the return is paltry. It's crucial to truly understand the differences in the nature of money received, and then carefully strategize about who to obtain money from and how to obtain it. I believe that finance is really all about designing relationships with the people who are the recipients of the money.

As an example of what Shizuoka Bank is doing in this space, I'd like to tell you about the recent trend of positive-impact finance.

An SME (small to medium-sized enterprise) wanted to borrow money from the bank to do something about business, but they were having trouble coming up with a story that would get their loan approved. They knew they should ride the SDG (sustainable development goals) wave and reduce their environmental impact, but they weren't really sure what to do. That was the situation.

In response, Shizuoka Bank suggested they develop initiatives for social impact. Are the human rights of foreign technical intern trainees in their workplaces being protected? What are the personnel systems, benefits, and workplace environments for protecting those rights? How could a system that minimizes CO_2 emissions from workplaces be designed?

Such initiatives can be measured using KPIs, but if the company is

Fig. 1-6. The structure of the world

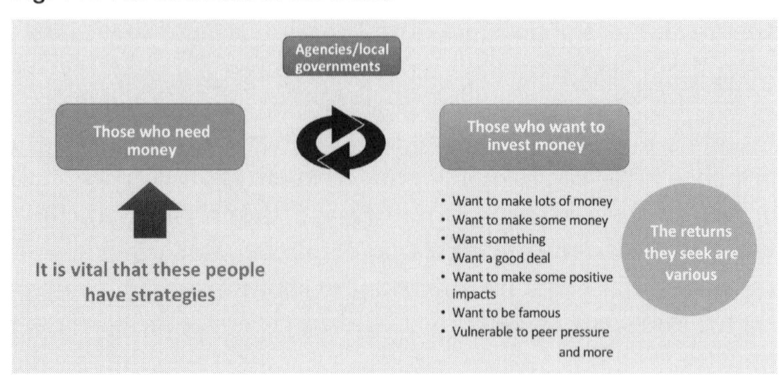

Fig. 1-7. Relation between money types and their returns

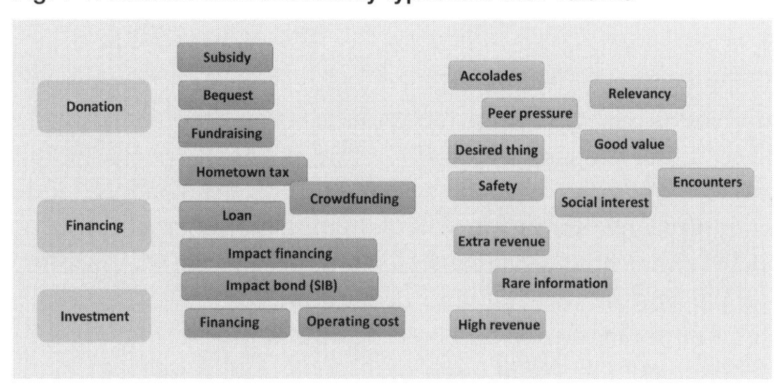

evaluating them on its own, it's no different from any other firm. So instead, they would get a rubber stamp from a third-party assessment organization, confirming that the environment they have created meets international standards. It would then immediately become evident to anyone that they are benefitting society, and those KPIs would be valuable for the first time.

Because they would have achieved KPIs whose value is recognized by society, not only would business improve, but the impact on society would be large. This gave Shizuoka Bank a legitimate reason to sign off on the loan, and as Mr. Abe mentioned, progress would also be monitored. This stimulated communication and enhanced disclosure.

That's the basic flow, but it's firmly in line with the approach to investment. Make a plan for the future, ensure it is taking you in the right direction, and devise indicators that can clearly demonstrate improvement. Then borrow money and make continuous reports as part of the monitoring process. By following this cycle, it becomes possible to secure funding that might otherwise be difficult to obtain.

I think it's vital for such a process to be designed in rural areas as well.

Flag bearers who take on risk and assume responsibility: General partners

Furuta: I believe there are various ways to invest in a region, such as constructing buildings, increasing the number of connected minds, and supplying knowledge. It's important, I think, to create modes of involvement that leave some form of asset in the region as a result of the investment. If that can be done, I think investments by companies could improve the economies of rural locales.

In that respect, I think the Hometown Tax Program is great, though it's unfortunate that the people who have donated money cannot take part directly in the initiatives of the region. The same goes for the corporate version of the Hometown Tax Program. They've gone to the trouble to invest, but the money ends up being spent without their involvement. If modes of involvement between companies and regions can be improved, I believe the number of companies looking to regions, like they do with Kamishihoro, could be increased.

When we talk about where responsibility resides, the presence of a general partner (GP), which Mr. Murakami talks about a lot, is also important. The GP is the person who ultimately bears full responsibility for the project. Without clarity on who this is, it is impossible to raise money in the first place. Someone must take on the GP role for each project, acting as the flag bearer in gathering money from investors. However, creativity is required in terms of how you make them ultimately responsible. The GP has to shoulder final responsibility themselves—a typical example is personally guaranteeing repayment of the loan. But even the finest GP is unable to simultaneously provide personal guarantees for multiple projects, so the result is that they can only handle one business. In regions in the future, when one project has gone well, several others connected to it will need to be implemented in order to create a chain of success. Forcing all responsibility onto the person in the GP role makes it impossible for that one person to simultaneously energize multiple projects.

Given this, I would like to see a guarantee mechanism, similar to special-purpose companies or social ventures, whereby the local businesses that are on board share the risk by each taking on a fraction of it, and credit is provided to the region as a whole. This could be similar to the positive-impact finance discussed earlier.

Abe: In the case of funds, the GP typically receives just over 2 percent of the money invested as a management fee, right? On top of that, they normally get around 20 percent of the investment profits as a success bonus. In the case of regional projects, however, it's clear that such bonuses cannot be expected, and with the risks remaining undefined, local governments might try to take advantage of the GP's services without compensating them adequately. I think the crux of the issue is what to do about this gap.

Essentially, it's important to accurately calculate the risk and design the return to reflect that risk. However, the reality for potential GPs in regions is being told, "We want you to take on unlimited risk, but because public funds are involved, your return will be limited." Nobody would want to invest on such terms, would they?

Murakami: This is important stuff, so let me summarize the key points. The GP has unlimited liability, and is the person who bears ultimate responsibility should the business fail. In contrast, a person who is only liable up to the amount they personally invested is called a limited partner (LP). Projects should combine the strengths of both.

There are two points. In the case of social ventures, one cannot expect much in terms of monetary returns, so the social impact becomes the return. For companies, this impact—namely, the contribution to resolving social issues—can lead to benefits such as increased sales of their products or a rise in their stock prices. This makes companies more willing to invest as LPs. The first point is how to create this structure. There is also another point: While monetary returns can be distributed, social impact cannot. Therefore, both the companies participating as LPs and the GP end up receiving the same thing. This leads to a situation where the GP is effectively undercompensated for taking on all the risk. As it stands, it's like asking the GP to offer unconditional love to the region. Clarifying the benefits for companies and eliminating the disadvantages for the GP are two significant challenges that will need to be addressed moving forward.

The inseparability of areas and projects

Abe: There's also the issue of how to define an area. An area is multilayered, so determining what constitutes a region depends on the extent to which the people inside it can exercise autonomy. It's difficult to objectively decide which part of this multilayered area should take ownership. However, the tricky part here is the administration. For local governments, the multilayered approach doesn't work; they need a clear target to function effectively.

Murakami: Our discussion has suddenly become more complex. Taking the example of the city of Mitoyo that Mr. Furuta introduced, there are cases where we need to consider areas that were delineated before municipal mergers took place, and cases where we look at a narrower area like the coast near Chichibugahama. On the other hand, from an economic or political perspective, all the areas would be categorized as being part of the city of Mitoyo, and if we think in terms of the economic zone, that could extend to include the city of Kan-onji; and collaboration could even cross the prefectural border to encompass the Toyo region.

Such differences of focus in areas need to be delineated somehow. If we don't limit our scope to a single area and focus on that, even if we want to organically link areas and businesses and get support, investors might become confused. Without clarity on which areas are connected and which information to look at, it will be impossible to generate excitement.

I believe the domains of the Edo period can offer a clue. In terms of both population size and geopolitics, I think the domain can serve as a criterion for establishing economically rational services. History shows that many domains survived independently for over a hundred years, which carries significant weight.

In practice, some businesses start in a smaller area and gradually expand out of it, while others operate in a large area from the beginning. In the latter case, I think it's fine for businesses to work with any local government they like, even a far-flung one like an enclave, to create services.

In any case, choosing which area to extract from a multilayered area structure means making rational selections based on such criteria as the potential for public services to be self-sustaining. It's a task that requires consideration of various aspects.

Takemoto: Area and social impact are two sides of the same coin, aren't they? Because the issue at stake is which area social impact should directed at, and who in that area should be affected. If it were in the realm of competition for independent financing, then in a sense, any area might do. However, given that it's about social impact, deciding where to draw the line between public aid and mutual aid, and in what area, becomes critically important. As populations continue to decline and markets shrink, if we continue only in the competitive realm without attempting to produce a co-creative realm, the people remaining in the regions will disappear, and eventually, everything will be lost.

Fujisawa: Hearing about the ongoing development of the new national spatial plan to create small "regional living spheres" with populations of around 100,000, as well as today's discussion, I wonder what the truly appropriate size for an area is.

Recently, I've been involved in the Japanese government's "Moonshot Goals," and we've been exploring the possibility of a shared infrastructure that allows everyone to participate in social activities through remotely controlled avatars. It reminded me that there's a need for shared infrastructure based on some form of mutual aid, and I wonder if there's a way for pioneering regions like Kamishihoro to initiate something and then expand it to other areas.

Also, a GP isn't someone you can just call up and bring in. I think we need a mechanism like that, one that makes people want to offer themselves up as a GP and come to the region of their own volition.

Miyase: Today's themes have been disclosure, governance, and finance, but when I actually asked the people in Kamishihoro, everyone, including the mayor, said that they'd been acting unconsciously. There might well be other municipalities out there that have already started creating mechanisms unconsciously, just like Kamishihoro has.

Session 5: Mr. Shintaro Kado's Case —Matsuyama, Ehime Prefecture

Wisdom transcending generations to bring the region together: The "Machi-Pay" circle of success, expanded through persuasion backed by data

Shintaro Kado of Machidukuri-Matsuyama left his job in Tokyo to return to his hometown and take over the family business; he has since vigorously devoted himself to realizing a regional recycle-oriented economy based in the local shopping district. During the 14 years of activity, there was drama under the surface as he battled with discord and conflict from traditional mentalities and gradually expanded the range of initiatives with relevant parties to foster a business that would serve as the core. With key phrases from Mr. Kado such as "data marketing" and "reverse-osmosis membrane," we trace the dynamism of how he engaged locals in a new challenge with the goal of rural revitalization.

A startup ecosystem for rural revitalization

Murakami: We sorted out the issues in the last session: The importance of a general partner (GP) who will take full responsibility for the rural revitalization project; how to foster the relationship between the GP and supporters while executing multiple initiatives, as if layering a mille-feuille; amidst those initiatives, how to establish an investment mindset in a region that is strongly oriented toward distribution for a shift to a business style that is fitting for regional development.

At this session, I would like to proceed in developing a template for successful planned rural revitalization so that those growth processes can be reproduced in other locations. This would be the creation of a startup ecosystem for rural revitalization, so to speak.

If one initiative that will serve as the nucleus is successful, can't we create an ecosystem by expanding upon that? Just what is the threshold for transition to that self-sustaining stage, and how do we cross it? And once it is crossed, how can we make that into a model and expand it horizontally outside the area? Using Mr. Kado's case of the city of Matsuyama this time, I would like to proceed with discussion from that perspective.

A general partner and area originating with a shopping district

Hori: Mr. Kado is a native of Matsuyama who graduated from university in Tokyo and went on to join the foreign financial institution Goldman Sachs. When his father became ill, Mr. Kado returned to Matsuyama and took over the family business, a clothing store. There, he also engaged in community-based businesses, including a restaurant and water-delivery business, among others. In 2014, his fifth year back in his hometown, he was appointed as chairman of the local shopping street. He is also currently the CEO of Machidukuri-Matsuyama Co., Ltd.

Kado: Yes, I am the third-generation head of Tokageya, which was founded 75 years ago. Machidukuri-Matsuyama existed before I returned to my hometown, but when I became CEO ten years ago, we took the opportunity to set the objective as "private sector-led autonomous community development" and have been engaged in regional city development, commerce promotion, environmental industry, and more. We started out with three employees, and we have nine employees today. If we include affiliates, I believe we have about 50 employees. This effort is funded by the city of Matsuyama, the Matsuyama Chamber of Commerce, the Iyotetsu Group, local financial institutions, and the shopping district.

Hori: That would be Matsuyama Chuo Shopping District, correct? We went to conduct research and found an arcade town about a kilometer long that comprises five areas, including Okaido and Gintengai. It has been a beloved area among locals for generations, and is used as a meaningful public gathering space. I heard that the shopping district has changed greatly with digitalization in the 13 years since Mr. Kado returned.

Kado: We launched a regional cashless payment service called "Machi-Pay." This combines electric payment, shared shopping points, gift certificates, and other perks that can be used in Matsuyama, enabling payment with a smartphone app or card. Shoppers can save or use points and receive coupons; these serve as incentives for locals to use

A scene from an event held on the major shopping streets in Matsuyama.

the service. At the same time, we use data from their purchases, as well as from tourism apps and footage from street cameras that capture town visitors, to illustrate the flow of people, money, and merchandise. This is returned to the shopping district as value and harnessed to trigger more activity. The locally operated infrastructure and data marketing system were recognized at the 2021 Innovation Net Awards, winning the Economy, Trade and Industry Minister's Award.

Hori: Does that mean, in a sense, that the company Machidukuri-Matsuyama played the role of a general partner for the specific area of a shopping district, using that as an originating point for amassing people and money while tying both to activities that would develop the region?

Kado: Yes, I believe that this type of town-development company can contribute to developing an urban structure that is site-integrated and site-coordinated. In terms of area, the scope for Matsuyama follows the Central City Area Activation Basic Plan, which covers the downtown area, the Matsuyama Station area, and the Dogo Onsen area.

Using data marketing to overcome competing interests

Hori: Machi-Pay, which was developed by Mr. Kado and his peers, was the centerpiece of the regional development measures. I heard that the issuance and promotional activities around gift certificates, for example, were not unified among the different shopping streets. But with the adoption of Machi-Pay, about 2,000 stores became affiliates, acquiring 70,000 members, to achieve efficient operations.

Kado: To set initiatives like these into motion, I think the mindset of making investments that will penetrate into the region is important. The regions are in a situation where they are competing against each other for pieces of a very small pie. To maintain governance while growing that pie, they need to share it, even with competitors. The idea is to use communication to ultimately develop a society that is free of stress.

Nevertheless, the process to this point was grueling. The area historically prospered as a castle town, so the locals each have their own expectations. To push the initiatives forward, we had to reconcile the interests of all kinds of people. With my background in the financial industry, I wanted to use data to reconcile interests and develop a consensus based on evidence. One successful example is the Ojoka Spring Festa.

For this event, we made the street connecting two shopping streets, Okaido and Gintengai, into a "pedestrians' paradise," closing it to cars so that families could enjoy themselves without worrying about safety. The objectives were to restore traffic that was declining in spots between the two zones and to enhance the connectivity and linkage between Okaido and Gintengai. We believed this would help to revitalize the overall shopping district, as well as the central area (called the Ojoka area). The event was held for the first time in March 2013 and continued annually until it was paused by the COVID-19 pandemic.

It may sound like things went smoothly, but there were hiccups along the way. It was 2011 when we first started discussions. Someone mentioned that you didn't see children in the shopping district anymore. I think most second-generation baby boomers would pile their families into their minivans and go to the shopping mall. Baby strollers and children became a rare sight in the shopping district. With a sense of crisis, 30 to 40 young people from some of the shopping streets, along with business owners, public administrators, and the chamber of commerce got together to consider branding for the town. The concept we verbalized and shared was "Aiming to become a fun and stylish town where citizens gather." It's ordinary as copy, but we made the decision to work together with a mission that no one could disagree with.

But once those initiatives were launched and I went out to explain, people, especially the elderly, responded, "Where's the money for that?" and "Using all that money, will it really work?" and even "Not over my dead body!" Talks went nowhere. And so I said, "Okay, I'll show you" and brought out the traffic-volume data.

I diligently went around and explained, saying:

> Please listen. The edges of these two shopping streets, spots A and B, have a fair volume of traffic. But spots 1 and 2 in the middle are very slow. Creating momentum here will increase overall traffic flow. To do this, please let us spend money to increase [visits by] families and create memories for children.

I took the data to companies as well. I was a novice business owner who had just taken over the family business. No matter where I went, I was the youngest, at age 28. I had neither money nor authority. The municipal office and police said no; after all, it's a major four-lane street that we wanted to make into a pedestrians' paradise. But at the very end, when we were just a week away and ready to give up, we were finally told, "Why not? Go ahead."

By consistently using evidence-based data marketing (assessment) to earn a reputation, it may be possible to change a region that is loose with money and lacks an investment mindset. I was able to sense this progress with the initiative.

A magnetic field created by an "anonymously made magnet"

Hori: A young man who had just returned to his hometown developed a large, unprecedented event in a world ruled by seniority. I imagine there were many challenges.

Kado: Yes, following precedent is the standard. Smartphones were not yet prevalent back then; simply telling people that a digital society would arrive didn't resonate with anyone. Showing concrete data was the first step to gaining understanding.

For example, what was the difference in the number of children elementary-school-aged and younger who came to the shopping district during the Spring Festa? It was about three times higher than the usual weekend. In the problem spots with low traffic, traffic increased by almost 30 times. By tenaciously collecting this data and persisting with activities while asking, "Wouldn't this make it acceptable?" people who were initially taking a watch-and-see attitude were gradually drawn in and joined the circle.

I think they were drawn in by what I call the three magic tools: an efficient urban structure, promotion of regional resource circulation, and acquisition of outside money. In addition, communication, coordination, and consensus were also contributing factors. This meant first creating an opportunity for everyone to participate, having all parties come to the table, and continuing to communicate across generations.

Hori: What do you mean exactly by "coming to the table"?

Kado: This does literally imply a roundtable discussion, but more than that, it is the concept of generating this gradually while ensuring that my presence doesn't stand out, hiding myself so it isn't clear who is the organizer. This may seem extreme, but there are times when people are like, "I won't go if he's there," or "Who is he to do this?" I am careful to avoid this, and will sometimes delegate the role or authority to others.

The result was an anonymously made magnet. They were impressively drawn in for some reason, and before they realized it, they were saying, "That's nice," "I'll do it," or "Let me do it."

Miyase: Is it unacceptable for that magnet to be made by Mr. Kado?

Kado: If one certain person is leader or director, people won't follow, because they are wary of being manipulated by someone. The same with Machi-Pay. Because I want everyone to participate, I don't want to say that I made it.

The key to drawing them in is to first create a place protected by "training wheels," where they can create a sense of relevance. In Mr. Furuta's words, that involves preparing roles to play and places to belong. I realized that if that fosters self-relevance, they will participate and come to the table of their own volition.

Hori: What do you mean by "training wheels that foster a sense of self-relevance"?

Kado: For example, there are the Ojoka cleanup activities. This is an initiative of the Institute Ojoka-Matsuyama, which I am chairman of. On the first Saturday of every even-numbered month, about a hundred people gather to pick up trash. We also remove graffiti. This has the effect of making participants feel virtuous. People who were carelessly tossing away cigarette stubs begin to subconsciously look at the ground and restrain themselves. As they pick up litter, the road becomes their own. In a sense, this also serves as training to participate in activities. That would be a simple way to put it.

A regional recycle-oriented economy, protected by a "reverse-osmosis membrane"

Hori: Most of the rural revitalization frontlines that we've visited up to this point seemed to have leaders who came from outside the region, and those inside cooperated to start something up. But in Mr. Kado's case, I sensed that the initiative was to encourage the participation of those with local roots, to cultivate the local people while investing in new fields—kind of like shifting to in-house manufacturing. Was workforce diversity also considered internally?

Kado: Not many people came from outside. My impression is that as we gradually expanded the circle of activity, diversity escalated progressively as well. Similar to how you have the town-development company, the shopping district association, the general incorporated association, and the areas, we proceeded while rearranging the different layers. This resulted in a slight variation of members according to situation and the amassing of human-resource layers.

At first, I was thinking about how to bring Matsuyama to the world stage. How can I make this town more abundant than anywhere else? How about sneaking in what is needed, only as much as needed, from big cities such as Tokyo and Osaka? Doing so as if surrounding it with a reverse-osmosis membrane—that was my approach.

Figuratively speaking, if there were cell membranes between cities and regions, these would normally function as osmosis membranes, siphoning off money and people from regions and transferring them to cities. Of course, plenty of funds flow from cities to regions as well, but even if these are applied to good causes in the regions, they ultimately seep out. I wanted to create a mechanism that would instead only adopt the benefits of cities and keep them inside the region.

In other words, this would be a system to draw in resources such as funds and personnel from outside while circulating those assets inside the region, doing as much as possible to prevent them from leaving. That would be "Machi-Pay," which I introduced earlier, and "comext," which we are now advancing. Comext is what could be called an operating system for the town, gathering and driving all services required for everyday living. We are aiming to develop this as an open platform of mutual aid that local businesses, individuals, and public administrators could freely utilize and integrate.

Miyase: Comparing Machi-Pay with a major cashless payment service like PayPay would make it easy to understand. If someone uses PayPay, even if they spend money in Matsuyama, the payment and data go to major entities. But if they use Machi-Pay, it limits the circulation of the data, money, and shopping points to inside the area. It's really the concept of feudal clan notes—the currency of the Matsuyama domain.

The issue of balancing osmosis membranes and reverse-osmosis membranes

Fujisawa: The balance of osmosis membranes and reverse-osmosis membranes is a key concept that really inspires a huge vision. The phenomenon of people, things, money, data, and all that seeping out may be occurring throughout Japan. If we discuss this carefully, it has the potential for widespread application. The thought is exciting.

Abe: This is intriguing. I love this topic of osmosis membranes and reverse-osmosis membranes. It's a dissipative structure, so to speak. That's my field of research. If a system is developed in a certain domain and remains completely closed off to others, it decays. You have to keep it just slightly open. But if it's too open, it collapses. Both osmosis membranes and reverse-osmosis membranes adjust this delicate balance. And this is not just for one factor; the issue is how to place cell membranes in the region for all kinds of layers—the flow of people, of money, of commerce, of information.

The layers touch each other, so wherever information gathers, so do people and money. There are points that link the market of information flow to the market of consumer flow. That leads to the question of where to create those nodes.

In the case of Matsuyama, the size is just right as well. It has a population of about 500,000, rather than several million or several tens of thousands, as well as cultural assets such as a castle and history. How is

the consumer flow developed in such a location? It's really interesting.
Maki: The comparison between PayPay and Machi-Pay really makes sense to me. For example, with the Hometown Tax Program coordination business, information migrates to operators that are well versed the realm of tax payments; it hardly ever reaches the local businesses. How is it possible to gain independence from those powerful platforms? In the region I am involved with as well, we are exploring systems that would enable us to circulate and accumulate information and money ourselves.
Furuta: Sapporo Drug Store in Hokkaido has a region-limited point card called Ezoca. They were invited at first by a major company to join their services, which was in fact tempting. But they decided instead to make one themselves. Apparently, they now have a membership of two million people.

With this type of system, it's important to launch usage all at once in many places. It's also attractive that it completely benefits Hokkaido. I think it succeeded because they insisted on using it to invigorate the local community.

A safety net to protect and incubate the next generation

Hori: There was mention earlier of human resources. How are you fostering workers on the front lines? When I went to research Machidukuri-Matsuyama, of which Mr. Kado is CEO, he said, "Employees joined this company because they thought the townspeople would be pleased, but in fact, the townspeople sometimes get angry at them. I feel that a safety net is essential to protect those personnel."

The city has different areas, each with shopping streets. The harsh reality is that they are presumably competing for customers. On top of that, there is a vague sense that town development is someone else's responsibility, and that the people or businesses that are engaging in it are subordinate to the shopping street itself. Under these circumstances, he felt he had to protect the young people who would become targets of criticism.
Kado: Yes. A few years ago, an employee said to me, "Are you going to keep this up? The people around you are not getting any happier. In fact, they're getting exhausted." And so I asked the founder of this company, whom I trust, to become the safety net.

Hori: That would be Jiro Hino—a strong ally who was chairperson of the Okaido shopping street for 20 years and had worked closely in partnership with Mr. Kado for reform. He has been involved in town development all around Japan since he was young. I also understand that he participated in the "Love Hachinohe" movement launched in 1975 by the Junior Chamber International Hachinohe of Aomori Prefecture. He was an advocate of the view that towns will not improve beyond what their residents are conscious of.
Kado: He is very passionate. He also says, "To know is to love." When I asked Mr. Hino, "What is the purpose of town development?" he answered simply, "It's for my own sake." I was taken aback for a moment. It's rare for a person to openly say that town development is for their own self-interest. But it was a revelation to me. In other words, it's "self-relevance." Everyone should have the mindset of relevance; instead of "someone will do it," everyone does it. That's why it becomes a roundtable, and that's why it creates a magnet.
Hori: This means that, until it gains that relevance, there is a need for a protective safety net and a mechanism like the "training wheels" mentioned earlier. In that case, what protects Mr. Kado himself? I asked an employee, and the response was, "The fact that he values what has been amassed in this region, such as its history, culture, and human sentiment." I also spoke with Mr. Kado's father, who said, "Shopping streets have an atmosphere of nostalgia and amusement that shopping centers are utterly lacking. I think my son understands that."

The shift to initiatives that layer like a mille-feuille

Takemoto: Hokkaido has an area called Niseko. Impressive hotels have been built there, and tourists come from all over the world. But actually, the action is not in the town of Niseko, but in Kutchan, the town next to it. Next to that is Rankoshi, which has no visitors—foreign or even Japanese. The issue is remarkably apparent when you look at these three towns in a row. Niseko, stuck in the middle, is really in conflict. The town wants to make efforts to preserve its uniqueness. But if they simply consider the quantitative aspects, following Kutchan's model would make sense. I think they need to talk it over properly if they want to put it into numbers. In that regard, I thought Mr. Kado did well in conveying the internal numerical element.

In terms of scale, an economic zone of 500,000 people sounds like a city, but Matsuyama is more like a town. Cities have a structure of anonymity, while in a town, the names of people, places, and organizations stand out. That is probably what generates the sense of nostalgia and amusement. The singular piquancy of a town would likely fade away with urbanization. Yet Matsuyama is taking the arduous and energy-consuming route, fighting to remain a town. I found that to be very moving.

Kamiyama: The home where I was born and grew up in was right by an arcade street, in a very traditional location called Juso near Umeda in Osaka. Because I grew up there, I know the ins and outs of a shopping district well. I can appreciate that Mr. Kado worked really hard.

In terms of the tourism industry, Mr. Kado's work is really the responsibility of the destination management organization. That would be to properly quantify the lifestyles in the region and the movement of visitors and apply the results to marketing measures. It's amazing in itself that he was able to do that in a city of 500,000.

I have one question for Mr. Kado—Can Machi-Pay points be exchanged at equal value for points, mileage, or electronic payments with other major services?

Kado: Actually, I talked to many people about that, but I didn't get much of anywhere. I was told that it made no sense to do that in a region.

Kamiyama: If you can do that, it'll make a big difference. The major services have made an ecosystem where they ultimately secure profit for themselves, so there is resistance to integration. You might consider telling them that this is not about keeping it in Matsuyama, it's about deploying a system similar to what other regions are doing, and that Matsuyama would just be the origin. You could justify it by saying that it would be a social contribution by the major services as well. I think it could work if you approached the right people.

Miyase: That is indeed a reverse-osmosis membrane. Mr. Murakami, what do you think?

Murakami: Creating a data-collaboration platform in the form of Machi-Pay in Matsuyama is really significant. I would like to engage in gradually developing open regional digital platforms like this in other regions as well.

A keyword to this, which was also mentioned at the start today by Mr. Abe, is "mille-feuille." Layering the initiatives in a certain area like a mille-feuille. There are proactive people in the area who come up with innovative ideas and energize those around them, and there are coordinators who level the environment. These entities work on multiple layers to generate activity, and when expansion is evident to a certain extent, they pause and freeze the movement. From there, they repeat their work to create something like a mille-feuille. That is how I imagine it.

In Mr. Kado's case, I think he has taken on too much of the role of the proactive person by himself. Normally, that individual would bring in some more colleagues from outside the region to work as a team, so there would be a competing force between the offense and defense. Mr. Kado has shouldered this on his own, so it was a huge struggle. One thing, though, could make the structural outline very clear. If we substitute a GP and an area for the combination of Mr. Kado and the shopping district, the significance of the "cell membrane" that exists there becomes obvious. I think that we can discern a formula from this that can be applied to other regions as well.

Abe: When we consider what is needed for each region to have a suitable economic zone, the case of Matsuyama is rich with implications. In terms of the osmosis membrane, Matsuyama itself can be the one that absorbs something from the surrounding regions of Shikoku. The area that spreads out with the castle in the middle facilitates the development of a nucleus for the whole, and there are multiple areas that can be enclosed in cell membranes. This all makes it easier to build a multi-tiered mille-feuille structure in a single town.

However, even if Machi-Pay functions like feudal clan notes in the Matsuyama domain, I'm not sure if that is suitable for an area of that scale. Listening to Mr. Kado and Mr. Kamiyama, I was thinking that when outside economic zones connect with the region without any exchange mechanism, the area with the greater economic power is definitely stronger. If, for example, I visited Matsuyama as a PayPay user, what would motivate me to commit to the Matsuyama economy, taking the time to exchange my collected PayPay points for Machi-Pay points that can only be used in Matsuyama? Considering that, I think that the key will be how Machi-Pay premiums are set and how consensus for integration is built in the overall region.

It may be possible to apply public funds to this. The question is how to justify it and whether it can be used like a regulator valve. In that sense, I am also watching the "Sarubobo Coin," which is a Hida area-limited digital currency that also links to government services, with great interest.

Giving voice to the silent majority

Maki: I don't mean to define Mr. Kado by his connections, but he himself is very interesting. While he has the elite background of having worked at Goldman Sachs, he is a friendly colleague in the shopping district; he is a person of logic, but of emotion as well. I think the community loves him because of that curious balance. And because he is working harder than anyone to be successor to the locality of Matsuyama, he can be an innovator. If we look at the perspective of what approach to take in engaging with regions, this has the potential to be generalized.

Hori: I agree that this may be a crucial point. He has the background story of coming home as the successor to a clothing store passed down for generations, but behind the humanism, he also has advanced financial knowledge. I sense that Mr. Kado himself is a "reverse-osmosis membrane" that draws in something from outside the region. I'm very curious if that could be reproduced.

Furuta: Actually, it may be impossible for someone who has never left their hometown. I don't think I could revitalize a shopping street in Roppongi, where I was raised. I honestly think it is amazing that he returned after having left and is now in this position, with the family business as his base. Seeing what he was doing, the older generation accepted Mr. Kado at some point. What triggered this change?

Kado: It was definitely after the success of the Spring Festa when everyone started praising me. I was told, "You make good on your promises. You're doing it for everyone's sake." I also had the opportunity to do a presentation in front of Minister Shigeru Ishiba (minister of state for special missions) in connection to a rural revitalization project, and that was one big reason the locals decided to accept me.

But even now, I receive no pay from the town-development company. Some people have suspected me of doing something shady behind

the scenes because of that, but I have stayed cool and ignored them.
Furuta: You're working for the sake of the region, so I personally think that you should properly take a fee.
Kado: The town-development company belongs to the shopping district, so I feel like working without pay can't be helped. I've even had a municipal worker ask me to do a job for free because I'm a local. It doesn't make sense, right? They would pay an external specialist if they were asking for their work. I think the attitude of cultivating people locally is weak.
Furuta: How about delegating everything to the region for a time? Then they'll probably realize, "It's definitely impossible without Mr. Kado." You could play that trump card.
Fujisawa: Hearing all this is making my heart ache. But I have a strong feeling that Mr. Kado has brilliantly created a silent majority. Would it be possible to use math to make the silent majority visible—to quantify them—rather than just to figure out the flow of people? Those with loud voices are usually actually the minority. I wonder if there is a system where the silent majority could communicate what they want to do.

Are you familiar with the tool called Decidim, a democratic platform for citizen participation that was developed in Spain? It's being used in the cities of Kakogawa and Yokohama. The locals raise issues directly, and people submit proposals as to how they would solve the issues themselves. I have a feeling that this strategy could also serve as a hint.
Kado: The DAO (decentralized autonomous organization) scheme that is gaining attention for the management and operation of blockchains is similar. The benefits of Machi-Pay have been surprising for the local people, too. If we're able to get more people on board for the next "comext" project, we may be able to use it to demonstrate our reach.

Creating a model of successful rural revitalization

Hori: Let's pause here and have Mr. Murakami organize the points.
Murakami: As part of the Cabinet Office initiatives to match local government issues with companies that can propose solutions, we held a workshop and collected opinions. Our results revealed the typical issues that regions need to resolve. This will overlap with the discussion here, but I will go over them now.

The first issue is the lack of a general partner. There are operators (limited partners or LP) who will invest with limited liability, but without a GP, governance cannot be developed completely. It is highly probable that the fundraising potential is not being fully explored.

The second issue is to energize both businesses and the area at the same time. This applies to Machi-Pay as well, but it is imperative to transcend the interests of multiple businesses, including competitors, to organize a single data-collaboration platform that can be used for a range of services and make this into a mutual-aid business model. No matter what the scale, there is need for a startup ecosystem that links an area with individual businesses, and investors, citizens, and businesses have to be rounded up as one team.

The third issue is delineation of the area. When promotion of a certain initiative in a certain region is done piecemeal by the individual areas forming that region, it diminishes the power of the message. An example could be the Echigo-Tsumari Art Triennale held in the Echigo-Tsumari region of Niigata Prefecture. A sculpture by contemporary artist Yayoi Kusama, which could be considered the event's symbol, is located near Matsudai Station. While it belongs to the town of Matsudai, it also belongs to the city of Tokamachi, and can also be considered an icon of the Uonuma area and the Snow Country Tourism Zone. If each of these entities promoted it arbitrarily, the power of this valuable tourism resource would be diluted. Thus, it is crucial to decide how to delineate the area.

Where and how do we establish this type of GP–business-area relationship? Unless a clear model is available, it will falter, even if attractive resources are available. Conversely, if a GP is identified, the area is delineated properly, and operations begin with an investment mindset that has a sense of responsibility toward funding, I think it would be possible to create a template for successful planned rural revitalization.

In fact, if we look at the successful cases of rural revitalization in various regions such as Mitoyo, Ama, Kamishihoro, Nishiawakura, Aizuwakamatsu, and Maebashi, some started by linking business and area, while others did so later. No matter what the order, I think it is possible to create a model for engaging relevant parties while getting closer to businesses, as long as the three issues are overcome.

The route to creating a start-up ecosystem

Based on all that, we gave some thought to the process for expanding the platform business mentioned earlier and forming a start-up ecosystem for rural revitalization.

First, when connecting municipalities and businesses, we should create lots of opportunities for people inside and outside the region to launch new projects. There still may be a loose attitude about money at this stage, but the aim is to hatch many "chicks" that have the potential to eventually grow into strong "chickens."

The next requirement is a space for incubating these into genuine projects. Mass creation of new projects is meaningless if they are not truly integral to the lifestyles of the region. This is the same as a tourism strategy being meaningless for an area without an understanding of the way local inns operate. If rebuilding and enhancing existing service businesses would benefit the residents of the region, integrating rural revitalization with government policies for the small- and medium-sized entities that are responsible for that would be a possibility as well.

Either way, it is important to cultivate spartan projects that won't waste funds or human resources, while linking businesses and areas.

The goal of rural revitalization is not to simply make the regions famous throughout Japan. That has no meaning unless the productivity and income of the local people improve. In other words, the goal is to revive Japan's economy. If we liken them to distal capillaries, local businesses are clogged. Unless blood flows through these veins, the overall health of the organism will not improve, no matter how hard the heart—which is Tokyo—works. This is what is happening with Japan's economy now.

So what, specifically, should be done? The first step would be to choose among various initiatives which project is key, and to focus commercialization support resources on that business while linking it to area strategies. A specialist team would then provide functions similar to what accelerators do in tech ventures, such as providing investor mediation, legal support, and business infrastructure.

When the key business has gained steady ground, digital platforms such as payment authentication would be used for horizontal deployment of the project. At the same time, efforts will be made for consistent messaging and discussion so that those engaged in the individual projects take ownership of town development, and also so that local people become involved.

The final step would be to expand the range of the initiative with the key business at the core, and to promote the attitude that money is meant to invest for the future. It would also involve further implementing digital platforms and fostering new key businesses while clarifying strengths unique to that region.

I hope that using this sequence of steps would gradually put all factors into place for the initial stage of rural revitalization. One additional point is that external support will probably also be needed for processes such as selecting the first key project, finding the key person, creating a network, and financing the expansion stage to take place after the project scale has grown to a certain extent. I think the presence of an area organizer who handles all this while coordinating support from many experts would be crucial.

A system for querying and responding to citizens

Abe: We have been engaged in the city of Tokamachi, as Mr. Murakami mentioned earlier, and are now on the verge of increasing our scale. As we inherit the initiative of the luminary Fram Kitagawa, who launched the Echigo-Tsumari Art Triennale about 20 years ago, how do we make this scalable? This would require tremendous cost, and I sense that we are in a very challenging position.

Hori: How has that been communicated to the citizens?

Abe: It has been broadcast gradually with the sentiment that we must maintain what Fram Kitagawa started. I think an important factor was that citizens were on the receiving end of media interviews. People coming from outside would ask locals to share their knowledge.

When being interviewed or asked questions, you have to say something, right? The process of transforming the Echigo-Tsumari Art Triennale into their own art festival by talking about it is imperative. It started out as a mysterious regional event, but as artists visited from abroad and people kept being asked all kinds of questions, it gained the position of an internationally known art event. The strong sense of citizenship was the key.

Citizens are prompted to think for the first time when people from outside ask them questions. As this keeps happenings over years and then decades, their feelings shift to "We want to preserve this region." In that sense, citizens do not exist from the start; they may gradually gain citizenship through engagement with outsiders and shift to take on responsibility.

For this to happen, it is essential to create a flow of people from outside who act as listeners. The key is how to design this.

Takemoto: That is true. I mentioned this before, but while people's autonomy and initiative are important, so is the action of responding. Citizenship manifests with that. I'm sure that the way and level at which people respond varies with the individual, so having variation in the prominence of citizenship could be considered healthy.

The first step of finding a key project

Fujisawa: I'm curious how the key project is identified. Deciding with a council system likely won't work. When a venture capitalist is searching for a project, usually one person chooses it and convinces the others while they scrutinize the proposal. And after the actual investment has been made, they put in all kinds of resources to accelerate progress. Does the area organizer mentioned earlier fill this type of role?

Murakami: I think that whether it is a company or a region, the same process generally would work. For companies, too, deciding the key project at founding is crucial. With the proficiency of the members of this Trailblazers Conference, for example, I think it would be clear which "chick" to choose. It is likely harder to find someone who would commit to taking on that project as GP.

As you say, developing a structure with the presence of an accelerator is also essential. But there is still almost no one in Japan who works as an accelerator, so assigning one per region would be unrealistic. For this reason, I think an organization is needed, like an acceleration team formed by experts.

Hori: In some cases, an issue or need within the region becomes the starting point of the key project; in other cases, someone from outside brings it to their attention because the local people aren't conscious of that issue or need. There may be several different approaches.

Fujisawa: A startup support program called the Antler Cohort Program recently came to Japan. It is a project with motivated accelerators gathered from around the world. They can select from a range of business plans and then actually support the startup. We may need a similar organization or gathering to choose key projects and GPs.

Training for perseverance and the power of one team

Kamiyama: There was mention earlier of creating lots of "chicks"; that is, launching lots of projects in the early stages. When this happens, I think how we shut them down is important, too. Generally, about 80 percent of new businesses fail. We must shut them down in a way that will minimize the damage. In terms of judo, this is falling, or taking *ukemi*. Judo beginners start by practicing *ukemi* before learning throws so they don't hit their head when they get thrown. It's the same thing.

Risk criteria should be set so that it's clear when to shut down a project if it doesn't go well after launching. Having this framework in advance would provide reassurance in moving forward. I think it's important to know, "We can go this far, so let's go on the offensive to that point." This would increase the number of "chicks."

In a sense, this is money education, although the local financial institutions should probably take on this part.

Abe: What he just said is really important. In terms of education, media has a large role as well. My company (Ridilover) strategically conducts "Social Issue Study Tours" for that reason. We started this because we think that hearing about regional issues secondhand is completely different from seeing them on site and in person.

About 10,000 people participate in this tour yearly, but if you ask, "Do those people go to the site and practice *ukemi* for developing projects?" the answer would be no. The number of people who operate the project would be a few dozen at the most, so I think we need training for this scale.

I feel that this should be done to some extent by public education. To improve Japan overall, we should have lots of people gain knowledge about regional issues at an early stage.

Murakami: Speaking of training, this may not sound proper, but I would like public subsidies to be used effectively. For example, even if someone is eager to launch a new business in their region, they may be hesitant because they only have experience in their family business and are afraid of ruining everything. I think there are a lot of cases like that. Considering this, they should use subsidies, or call for investments and launch a separate company from the start. I would like them to use that idea to ambitiously take on all kinds of new challenges.

However, it is also preferable at that stage to have a key person even remotely available for town development in that area. If that person, as an advocate, would keep doggedly flying the flag for as long as it takes, it would make it easier to persist with the different initiatives.

There are likely many models, with that person being a GP, an accelerator, or an area coordinator. But ultimately, I think it is important for all parties to form one team and devote themselves to rural revitalization together.

Session 6: Mr. Toshiki Abe's Case —Echigo-Tsumari, Niigata Prefecture

Creating a "kelp forest" of human resources and businesses: A rural revitalization ecosystem emerging from an art festival

The region featured in this case study is Echigo-Tsumari, an area in southern Niigata Prefecture which includes the city of Tokamachi and the town of Tsunan. The Echigo-Tsumari Art Triennale, an art festival that has been held in this majestic natural setting for more than 20 years, attracts many tourists from around the world. This time, we will examine how efforts to build up a mille-feuille-like initiative with key businesses at the core have been nurtured in this region, where many of the elements of the discussions at past sessions of Trailblazers Conference are concentrated. We will also examine the potential to expand horizontal development. This case study was related by Toshiki Abe (Ridilover Inc.).

A company that breaks through social indifference

Abe: First of all, let me explain what we do at Ridilover. Ridilover started as a volunteer organization 14 years ago with the aim of breaking down social indifference. It has since grown into an organization that operates under the slogan, "Social issues for everyone." More often than not, social issues cannot be solved by the parties they affect. So they become problems for society, but that doesn't mean that others are very interested in them. I didn't feel that this was going to solve these problems and so my motivation for starting Ridilover was to do something about this contradiction.

How should we go about breaking through this indifference? Please look at what we call our "enthusiasm map" (see figure 1-8). This figure represents the transformation by one person who comes into contact with a social problem, understands it, and becomes a solution leader. As we move toward the center of this circle, the number of people marked as "impacters" in the middle increases, thereby involving society as a whole. That is what Ridilover does.

Fig. 1-8. "Enthusiasm map" and business objective of Ridilover

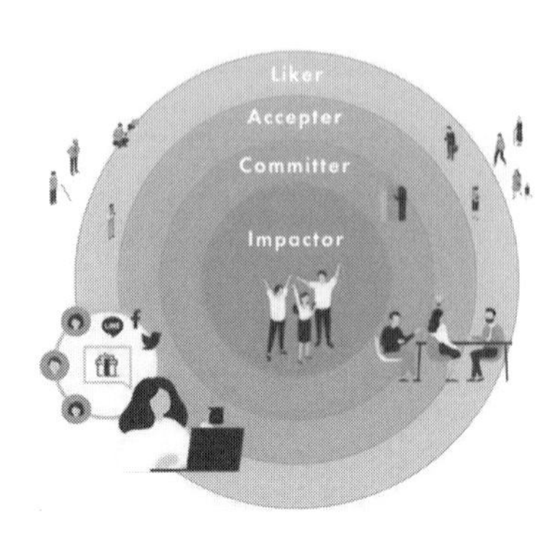

Liker
People basis
They have met someone involved in a social issue and have become involved themselves
Accepter
Issue basis
They are knowledgeable about a particular issue and can explain the background
Committer
Action basis
They are using their own time to work on solving the issue
Impacter
Results basis
They are promoting real problem-solving

On the outside of the circle, for example, are "likers." Even if a person is not well versed in a certain social issue, they can develop an interest in the issue when they meet people involved in it. Getting them to know about an issue on a people basis is the first step.

The second step is to make the liker more knowledgeable about the issue. This is an issue-based step during which the liker becomes an "accepter" by investing time in understanding the problem.

The third step is the "committer" step. During this step, we get the person to actually take action and spend their own time and money to work on the issue. This includes donations.

Finally, the last step is the "impacter" step. We want to increase the number of people capable of promoting problem-solving not only by themselves but also by involving others. For example, if you start a non-profit organization (NPO) that works on solving problems, you can create jobs, which links to the emergence of new committers.

Our mission is to make this circle bigger and bigger while increasing the number of impacters who attract more and more people on the outside. Consequently, a high-priority key performance indicator (KPI) for Ridilover is not only money, but the number of people spending time and the amount of time spent on a particular social issue.

Creating an ecosystem for solving social issues

These days, social problems are becoming increasingly complex and are taking an extremely long time to solve. More and more people are crying themselves to sleep because they cannot solve their problems. I want to figure this out as soon as possible. To this end, Ridilover operates in three stages.

First, we work to identify the issues associated with someone's problems. Second, we work to make as many people as possible aware of and involved in the issues we have identified. We call this "socialization." Third, we invest resources. "You all are aware of this problem, right? Let's solve it!" We get the people we have involved at the socialization step to contribute their expertise, money, and other assets.

What are we doing in practice? For example, during the issue identification stage, we report on our research, which is published in the *Ridilover Journal.* During the socialization stage, we organize study tours for junior high and high school students and corporate training courses, based on our reporting. We take people to see social issues with their own eyes and learn more about them. The number of participants in our study tours is currently averaging around 10,000 per year. To learn about food insecurity and food waste, we have taken participants to homeless sites and shops. Such visits are often incorporated into longer school trips.

Then, during the resource-investment stage, we formulate public policies and make proposals, linking them to test projects, and we keep pace with the development of new businesses by corporations. Recently, we were involved in a project that managed a 10-billion-yen impact fund that invested in listed companies while measuring their efforts to address social issues.

In short, our job is to identify issues, make society aware of them, and pave the way for everyone to work together to solve them. In other words, we want to create an ecosystem for solving social issues. Generate interest, train people, introduce projects, create systems. It is essential that we create systems and then return to the task of generating even more new interest. Otherwise, the ecosystem will not work. This is why we continue to involve people at every turn, regardless of background, whether they be students, citizens, corporations, or government employees.

Places where social issues are becoming more and more apparent

I will explain the differences between urban and rural social issues with

Fig. 1-9. Issue solution approach map

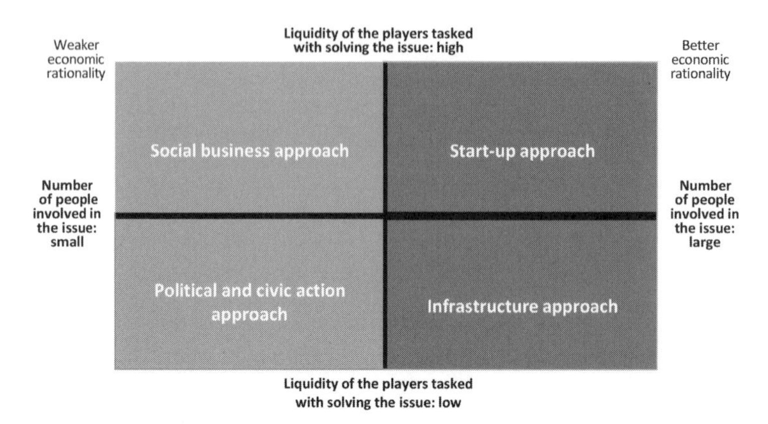

a matrix that divides the differences into four categories. The x-axis is the number of people involved in the issue and the y-axis is the liquidity of the players tasked with solving the issue.

At the top right there are both many people affected by the issue and a large pool of players who can get involved in providing solutions. These kinds of issues are better suited to being solved in the so-called startup world. Take digital transformation (DX) in urban areas, for example. As is apparent from food delivery services and taxi apps, there are numerous examples in which Software as a Service (SaaS) startups can provide business solutions.

On the other hand, there are cases such as energy issues in which the number of people affected is high but the liquidity of players tasked with solving the issue is low. Since energy projects become infrastructure projects, a large corporate approach is appropriate. Conversely, a social business approach is better when there is some liquidity of players tasked with solving the issue but a limited number of people involved.

Finally, at the bottom left, the number of people involved is small and there are few people available to solve the issue. In such situations a political, governmental, or civic-action approach is the only option.

This fourth area is where social issues tend to manifest themselves. In urban areas, such issues include child poverty, LGBTQ issues, social withdrawal, and homelessness. In rural areas, themes that should be approached from a business perspective in urban areas may end up becoming social issues. Take the DX example I mentioned earlier, for instance. There are only a handful of regionally based small and medium-sized enterprises with a reasonable budget to tackle this issue, and there are few experts who can solve the problem. We may no longer be in a situation in which businesses have a foothold on the social issue. In this case, the government will be the responsible for finding a solution. As a result, the government is becoming the largest player in solving issues at the local level.

This is how various issues that originated in the business realm are becoming social issues in regional areas, and this is the situation we acknowledge first when we identify an issue.

Developing connections through the Echigo-Tsumari Art Triennale

Let's move on to the main topic. I am going to talk about our activities in the Echigo-Tsumari region of Niigata Prefecture. Echigo-Tsumari, a mountainous region that includes the city of Tokamachi and the town of Tsunan, has an aging population and many marginal settlements. The area is known for having one of heaviest snowfalls in Japan; in winter the first floor of the houses there are completely buried under snow. Shoveling snow is extremely difficult, and people dying due to a fall from a roof while shoveling has become a social issue. Our first encounter with Echigo-Tsumari was actually totally unrelated to the art festival. It occurred more than 10 years ago, when we visited the area on a snow-shoveling tour.

The Echigo-Tsumari Art Triennale was first held at this location in 2000. The Triennale, held once every three years, has continued as a contemporary art festival for more than 20 years, and has grown into one of the largest international art festivals in the world. In addition to the Triennale, about 200 other artworks are exhibited throughout the year in the rice paddies in the sprawling *satoyama* area between foothills and flat land, which is used for agriculture and forestry.

At first there was strong local resistance to installing art in the rice paddies, and so the people involved in the project started farming as a means of building relationships with the local community. Production of the popular Koshihikari variety of rice thrives in this area, but there were an increasing number of terraced rice paddies and abandoned

More than 200 artworks are exhibited in the rice paddies of Echigo-Tsumari all year round.

fields that the locals were unable to care for. So the organizers established an NPO, the Echigo-Tsumari Satoyama Collaborative Organization. By working on the farmland, they were gradually able to build up trust with the local community members who, in return, began allowing them to install artworks.

It was locally born art director Fram Kitagawa who first proposed the idea of holding an art festival here. Artworks are considered easy to trade as a financial asset when installed in urban areas. Mr. Kitagawa felt that this was speculative, and not what art should be. He wanted to install artworks in a rural community to make them "fixed" property that could not be traded. That was the beginning of the Echigo-Tsumari Art Triennale.

Now, when a Triennale is held, it attracts more than 500,000 visitors to an area with a population of just over 60,000, with a reported economic impact of five billion yen. If an event with an analogous economic impact were to be held in Tokyo, it would translate to about 600 billion yen. This is a tremendous impact. Naturally, there are complex relationships and difficulties involved in organizing such an event.

Ridilover was not involved in the art festival from the beginning. We got involved six or seven years ago. Kohey Takashima, CEO of Oisix, decided to set up an organization of "official supporters" as an advisory body for the art festival. When Mr. Takashima asked journalist Daisuke Tsuda if he knew anyone who could put together a guided tour, Mr. Tsuda told him about "a guy named Abe," and I agreed to help. This is how we got involved.

The Echigo-Tsumari Satoyama Collaborative Organization is the organizing body, so to speak, behind the Echigo-Tsumari Art Triennale. Staff members, both those born locally and those who have moved into the area from within and outside the prefecture, work together to manage the artworks and facilities, organize exhibitions and workshops, and conduct public relations and activities to attract visitors. How this organizing body is run is critical. Although Fram chairs the organization and Mr. Takashima acts as vice chair, behind the scenes, various companies are involved as official supporters. Ridilover's commitment is particularly high. By communicating with local governments on other projects, we sometimes act as a bridge between the NPO and them.

We are increasingly using Echigo-Tsumari as the venue for our corporate training programs, and if you include our other projects there, we are investing a lot of effort into our activities in this area. To date, we have been involved in activities in various regions, and we have found that there are some things that we cannot accomplish by operating in a decentralized manner. We thought that we might see something different if we concentrated our resources in one area.

When promoting our activities, I believe that how we go about creating a program that effortlessly and continuously increases community involvement is the key. We need to make our activities more accessible to increase exposure and to get more people interested in them. Then we need to find ways to get those people to actually come to the area and get involved. We put together programs for this purpose, and encourage people to participate just once at the start, and then on a regular basis. While organizing such programs, they gradually found themselves involved on the project creation side. Once a project achieves results, it is then their turn to start involving others. We are intentionally working to create such a cycle.

Systematizing the function of the general partner

The first topic of today's discussion is where general partners (GPs), the unlimited-liability investors who run projects, come from. The second topic is how to make the series of programs focusing on the Echigo-Tsumari Art Triennale into a sustainable system.

First, I see GPs as asset managers for investments in the community. Their job is to extract money from people, secure resources, and find projects to invest in. I think resources should encompass not only money, but also human resources and people's interests.

However, when it comes to finding and carrying out projects, there are virtually no projects in rural areas to invest in. Therefore, GPs have no choice but to come up with their own. But even if they do so, they cannot expect a return. In regions with declining populations in particular, an economic impact will not be generated, and so the returns are small compared to the risks involved. This makes it much more difficult for GPs to secure resources.

This vicious cycle is the structural reason why it is difficult to cultivate GPs in rural areas and why rural revitalization is not progressing.

Hori: The role of such GPs seems to overlap in part with the farm stay, castle stay, and *albergo diffuso* initiatives that Mr. Kamiyama is involved in.

Kamiyama: My approach is different, but I think there is some resonance in terms of creating demand in the region. Mr. Abe is creating a new need, the need to solve social issues, while running the tourism and event aspect of his business. His approach is like a new kind of conscious travel agency.

Abe: Yes. Up until now the travel industry made its money by offering to show people something unusual, but these days people can go and see those things for themselves on their own. So the travel industry has to start by going to the locality and creating new demand. That is what Mr. Kamiyama has been doing. For us, the attraction we offer is solving a social problem.

When it comes to GPs, I think it is more important to create a system that guarantees their role as a function rather than having one superman set up a project. If they take on the challenge by themselves and fail, they will not get another chance. To avoid this, GPs need to diversify risk, visualize returns, and measure impacts. I think that systematizing the function of the GP in this way is Ridilover's role in rural revitalization.

Furuta: That is an important point. In reality, people who take on challenges tend to go it alone. I would like to add that the function of GPs is not only to launch projects, but also to create a platform in which local businesses can participate. That said, GPs are not just about opening a shop or whatever. They need to create a platform, get everyone on it, and take responsibility for it. Essentially, it should be possible to create a model that functions as a system, but unfortunately this is not happening.
Abe: The issue is whether the GP function can succeed as a business. If a situation is created in which KPIs are clear, and in which they will grow and yield profit with enhancement of the GP function, then the GP function will take on a life of its own.

Kelp forests: Creating GPs and training stakeholders

Let's make an analogy in which GPs, the entrepreneurs they bring with them, and the people working to improve the region are like fish. If we want fish to thrive, it makes no sense to just bring them together. In order to facilitate their growth at an individual level, an environment like a kelp forest is needed. We've talked about this before at a previous session.

If I were to try to explain what kind of place this kelp forest is, I would say that it's a place where a sense of security and incentives to take on challenges coexist. Fish eggs will not start growing just by being released into the sea. Only when there is kelp, oxygen, and plankton can an ecosystem be created in which fish can emerge from their eggs and grow up safely. What is important is not only the fish, but also the kelp forest. Having such an environment ensures that mediators—in other words, GPs—will be present to give young fish encouragement while providing them with opportunities to take on challenges.

Another aspect is the incentive to take on such challenges. We believe that this requires the existence of onlookers of a sort. A GP left to his or her own devices will not be able to produce results or keep working hard for long. It is only when there are spectators watching the GP struggle and praising him or her by saying, "That's good," or "That's great," that the GP can continue to work hard. By spectators, I mean interest from society.
Fujisawa: I feel that this kelp forest is Ridilover. But, when you put it

that way, it sounds like the incubation mechanism used to produce entrepreneurs. So I was somewhat curious as to why you are talking about a kelp forest, not an incubator.
Abe: Incubation is probably the way to create the top tier. Incubation finds a group of promising people and makes them compete with each other to create star talent. But regions cannot function with just top-tier talent. When a person who is incredibly sharp and astute enters a region, they could be crushed. In fact, someone who can't do something without help from someone else might be more suitable.
Fujisawa: In that case, I feel that kelp forests need other elements. Elements that can enable people to grow quietly without being made to compete, as Mr. Abe has just described.
Abe: I see. That may be an "interest-building" element. How do we find spectators and how do we increase followership? I feel this will be one of the key points.
Maki: My impression of a kelp forest is that it must be an amiable place. It must make fish feel that they belong so that they can grow at their own pace, hiding in the kelp, without being defeated by the big fish outside.
Hori: This concept seems to have something in common with the "safety net" Mr. Kato mentioned at the last session.
Kato: I think the important parts are probably the same. In addition to having enough oxygen and plankton, it is also important to be hidden from and invisible to others, and to be unknown.
Abe: It's true that being able to hide is an element that links to the mille-feuille-like multilayered structure. I think of it as a multilayered community with places to hide, places to belong, and fellowship. An environment that is not just harsh.

Multilayering with internal and external interaction

Let's talk about the Echigo-Tsumari Art Triennale as an example of multilayering. When Fram Kitagawa first suggested holding an art festival here, his role was as a GP as well as an area organizer. At that time, Mitsu Hara, Fram's nephew, joined him from Tokyo, and together they began preparing for and explaining the art festival to the local community. Fram played a central role in communicating with the locals, while Mr. Hara kept pace with him and gradually evolved from a city person to an Echigo-Tsumari person.

At the same time, people who were not necessarily in favor of the art festival gradually became more open to the idea, and more and more people began to work as volunteer staff or on the organizing side. This change was not just brought about by the local community, but was the result of an interactive process encouraged by the addition of people who came from outside the area.

Now, 20 years later, these relationships have become very important. While Fram is still responsible for bringing in artists from around the world and raising funds from outside the area as a GP, the role of area organizer, who is responsible for organizing the community, is now in the hands of Mr. Hara. Other people who act like area organizers have begun to emerge from within the local community, and the project is now becoming more and more multilayered.

So how are we get involved? We work together with other official supporters on aspects that Fram and Mr. Hara cannot cover by themselves, and on bringing in outside resources in particular. In Ridilover's case, we do not bring tourists, we bring people who intend to take some kind of concrete action, such as learning about local issues, starting a business, or moving to the area. And as we got involved locally, we created a situation in which were gradually drawn in and got more involved in Echigo-Tsumari. That is what makes us unique.

In fact, there are people who have made a career move into our company after participating in our study tours or corporate training programs, and there are also people who have moved to the area and are working for the Echigo-Tsumari Satoyama Cooperative Organization.

I believe that bulking up the layers between the people on the inside who are close to people on the outside and people on the outside who are close to people on the inside is essential for the revitalization of the region. That's how we see the role of kelp forests.

Kamiyama: I see. So you are saying that the Echigo-Tsumari Art Triennale is able to continue because of this approach. But I expect the people who are involved now will eventually leave. I am still very keen to know what mechanism will be used to ensure the Triennale continues after that.

Unified by a love of community

Hori: Ms. Miyase and I were also able to visit Echigo-Tsumari during the recent Echigo-Tsumari Art Triennale. It was there that we met Mitsu Hara, the area organizer Mr. Abe mentioned. Mr. Hara is the executive director of the Echigo-Tsumari Satoyama Collaborative Organization. Mr. Hara told us that, looking back on past 20 years since the festival was launched, it was not exactly easy to gain the understanding of the local community.

> It was really hard to get the project off the ground from scratch and with no one understanding what we were trying to do. Everyone scorned us. The art world, the local community, everyone. That was because we were trying to scatter 200 or 300 works of art on land the size of Tokyo's 23 wards. Nevertheless, when we saw the numerous issues left unaddressed in the area, such as vacant houses, abandoned schools, and rice paddies, we decided we would do everything ourselves, despite having no previous experience. As we took on the work, things gradually changed. We made a lot of mistakes, but we kept on going.

Initially, Mr. Hara and a group of university students from the city went door to door visiting each and every local home to explain the purpose of the art festival. Since it was also a public project using taxes, there was some opposition from the local council.

As they persevered, some people sympathized with their stance and took on coordinating roles on the local side. One such person was Shizuko Mizuochi, who has been involved in community development centered on "Ubusuna House," one of the festival's artworks, and who now serves as a city council member.

> The local residents trust the NPO, but still feel like they are outsiders. It is obvious that Mr. Hara and his staff are working very hard to connect with the locals, and not just to ensure that those who come to the Triennial have a good time. So I asked myself what we could do, and I took on a coordinating role.

These relationships of trust with the local people have supported the management of the art festival. So what are the changes we will be seeking in the future? Mr. Hara said, "If more people, especially corporations, get involved from a public-interest perspective, I feel that we will be able to get more things done, not only here, but in many other places."

Miyase: What stood out to me in what Mr. Hara said was that their 20-year effort has taken root precisely because they built relationships in which they were able to talk with the local people honestly about how much they struggled to get through the toughest snow-shoveling season in the area together. Without something like that, I feel that it would be quite difficult to replicate this kind of initiative in other areas.

Hori: Mr. Hara has family in Tokyo, so his base is Tokyo, but his roots are in Niigata. He had memories of visiting his grandmother's house when he was a child, and he wanted to get involved in this region in some way. I think that this is another point that has to do with reproducibility.

Miyase: I agree. If you have some kind of emotional connection to an area, you may be able to keep your project going for a long time. That said, a person like Mr. Hara cannot take on everything. So, assuming that people with ties to the area are at the center of a project and are deeply committed to the area, I am wondering how you would go about providing the complementary functions that support them.

Fujisawa: I thought Kohey Takashima, an official supporter, was a key person. He was a high school classmate of Mitsu Hara. Mr. Takashima's role in bringing people and money to the project is quite significant. As is Mr. Hara's managerial support and emotional backup.

Miyase: With the participation of the corporate supporters that Mr. Takashima approached, it would seem that various new initiatives are underway. I think this relates to the public-interest perspective that Mr. Hara mentioned. If more corporations consider investing not just for economic benefits, but also from the larger perspective of improving the well-being of the community, I think we will see more new initiatives.

A reproducible template for rural revitalization

Abe: There is certainly an emotional aspect, but in fact, behind the scenes at Echigo-Tsumari, a conscious effort is being made to create a "mille-feuille." The reason for setting up an official support system is one example of this. Fram is a well-respected person in the industry, so it is difficult for some staff, including Mr. Hara, to assert their opinions. Discussions come to a standstill in such situations. Mr. Takashima was therefore invited in an advisory capacity to help create an environment in which discussions could be facilitated. A number of people, including us, were added to this process.

Management discussions are also held with Fram and Mr. Hara once every two months in Tokyo. Topics include, for example, how to increase the number of tours with revenue potential. Being able to make such management decisions is a major change that has come out of Mr. Takashima's participation.

Recently Ridilover launched the "Closing the Experience Gap for Children" project. The project brings children from struggling families to the Echigo-Tsumari Art Triennale area to give them an opportunity to learn through a range of experiences. With an annual budget of tens of millions of yen, it is another type of funding initiative to bring people to Echigo-Tsumari. These are the kinds of projects we are working together on.

Personally, I would like to create a system whereby indirect financial resources for the Echigo-Tsumari Art Triennale can be covered by donations from individuals, while these kinds of projects continue to be introduced. In fact, the marketing of donations to NPOs has come quite a long way. I would like to see such a system used to defray the cost of preserving the artworks and on personnel expenses.

Maki: I am sure that Mr. Hara struggled a great deal. But people like Mr. Hara are very much needed in the field. I think that Mr. Abe has created a system to prevent such people from being worn down. It's a system of support from surrounding people. I feel that, ultimately, a "kelp forest" is a place to receive support from the outside while defending against attacks from the inside.

This overlaps with what Mr. Furuta just said about platforms, but by creating a system that encourages open participation and disperses risk, it is possible to both gather support from the outside and receive emotional support. On the other hand, we still have systems that reconcile interests and share out profits within regions, so the old structure and the new structure must coexist while constantly competing against each

other. That is why support from within can become stronger when it is inspired by support from outside. That's what I have concluded.
Abe: That is exactly right. I think the secret to why people are drawn to Mitsu Hara lies in his strength to say, "I'm going to make it happen, even if it's not efficient." In fact, there are times when it is better to do something inefficient that increases social interest and makes it easier to attract various resources. I am sure Mr. Hara understands this strategy.

What I mean by this is that, for example, in winter, a fireworks display called "Yuki Hanabi" is held on a snow-covered field. Local elderly people are involved in preparing for this, as volunteers, by burying 30,000 LEDs in the snow. This is an extremely challenging event in terms of profitability, but now, after organizing it for quite a few years, it is finally making a profit. This has led the locals to suggest adding more fireworks next time. The natural first reaction is, "Why? When we have finally started to make a profit?" But it is precisely because these people are willing to work without regard to profitability that we feel compelled to support them from the sidelines. No one really likes a community that is only ever thinking about making money. Mr. Hara seems to understand this, and acts from a slightly broader perspective.

We share Mr. Hara's advanced strategic thinking and encourage him to be as inefficient. That way we can increase involvement from outside. Interest from outside will lead then to a change in attitude on the inside. This is partly possible because Mr. Hara's perspective is amazing, but on the other hand, I also think other areas can have this perspective too.

I think the proof that the initiatives in Echigo-Tsumari are a mille-feuille is the fact that we talk about the Echigo-Tsumari Art Triennale as if it were our own project. I did not start the Triennale. People who had nothing to do with us started it, but they eventually formed relationships, successfully created a system to generate profit, and oversaw the development of numerous projects.

After all, what we believe is important is to create a reproducible template for rural revitalization. In other words, the question is how to create a cycle in which, after an individual project is launched, that project generates the next round of interest. First, we need to create projects. That project will create the next spectators. The next star will emerge thanks to those spectators, and this will lead to the inception of the next project. It would be nice if this kind of circle could be replicated in other regions.

The "inverse T-shaped model" of rural revitalization

Murakami: I'd now like to summarize the efforts required to create an ecosystem for rural revitalization. There must be many ways to reach this objective, but to keep things simple, I will use a simple structure, the inverse T-shaped model.

First, the vertical part of the inverted T is the "key business" that we discussed in our last session. The point is to choose a key business that is likely to have a ripple effect on other projects and that incorporates good governance, disclosure, and financing. And then to create a digital platform that utilizes My Number Cards and so on. This is the horizontal part of the T that supports the key business. And that is why it is an inverted T.

Let's take the town of Taki in Mie Prefecture as an example. Vison, one of the largest commercial resorts in Japan, is located here. It attracts more than 10,000 visitors a day on average, but only a few hundred people can stay in the adjacent accommodation facility. This being the case, the Vison brand can be used to send customers to nearby accommodations. In this case, accommodation collaboration is the key business that will strengthen relationships between accommodation businesses by creating a payment and authentication system using My Number Cards, and by providing discounts and points through cashless payments.

This creates an inverted T. Once this mechanism starts to work smoothly, the next step is to involve local specialty products and restaurants and incorporate popular stand-up paddle board experience tours to come up with affordable plans for sightseeing the locality that include everything visitors need. This is the kind of development that can be envisaged.

The important point is to create a tight model for the key business so that it does not lose out to local pressure to distribute, and to gradually bring surrounding businesses over to the "investment-culture" side while leveraging the synergistic effects of digital platforms. Then, mille-feuille layers will be added by incorporating various elements such as experts and funds from outside the region while gradually increasing the pillars of the business.

The lead-up to the creation of a rural revitalization ecosystem

So the question is how to proceed. Let's review the whole process based on our previous discussions.

The first step to encourage a local sense of urgency. Data can also be used to convey the reality of the situation. After that, the next step is to bring together many entrepreneurs who are just starting out or are looking to start something new. As was mentioned during the discussion of "kelp forests," adequate safeguards should be provided while also creating an environment in which startups can have a go at any time. This is the preparation stage.

The next stage is to work on a key business. There are three important considerations for running a successful enterprise. The first is distance from the municipality. If you get too close to the municipality, you risk succumbing to the distribution culture, while a business that is not recognized by the municipality risks not being successful. It is essential to maintain an appropriate distance from the municipality and the municipal council while also skillfully cooperating with the head of that municipality.

The second consideration is social impact. This refers to giving the key business an easy-to-understand cause. Clearly articulating this cause will make it easier to expand the circle of projects and get people interested in the next project.

The third consideration is financing: creating appropriate relationships with people willing to give money. Relationship building can manifest in a variety of ways, including donations, loans, and investments. Looked at from another angle, this means establishing governance.

It is impossible for a general partner to try to do these things alone. An accelerator who keeps pace with the GP is needed. That said, it is difficult to create accelerator teams for each region, and so organizing an all-Japan accelerator team that transcends areas is something to consider.

At the same time, a commercial structure that involves local citizens and businesses needs to be created. However, it is important to expand efforts gradually, like stacking the layers of mille-feuille, one by one.

Ideally, each layer should have a set of people who take on new challenges and people who mediate the conflicts that arise.

As the project evolves in this way, people on the inside, who are steeped in the distribution culture, are lured to the investment-culture side. On the other hand, people with an investment-culture background who came to the region from the outside are provided with a "fair environment," in a different sense from conventional fairness. It is not fairness in the sense of doling out subsidies evenly, but fairness that promises free competition based on certain rules.

This is how the mille-feuille becomes multilayered. Then comes the area theory. In other words, rolling out services to areas that are not restricted to the area administered by the municipality. Private-sector businesses led by local governments inevitably limit their service areas to administrative areas. How do we overcome this? It is not simply a matter of where to set the boundaries, but an attempt to change the meaning of an area from a distribution-culture area to an investment-culture market.

The final stage is the use of open data and digital platforms. As Mr. Kado mentioned during the previous session, the distribution-culture stronghold can only be broken with concrete data, not "should" arguments. The quality and low cost of digital platforms should be an important accelerator in expanding the circle of initiatives starting with the key business.

Will accelerators really show up?

Hori: You mentioned that the need for accelerators. How can we create a situation like the one in Echigo-Tsumari, which has Ridilover and Oisix and people like Daisuke Tsuda?

Furuta: I think various people can come into a region. There are people who come up with ideas, people who bring in capital, people who can connect government and businesses, and people with various specialties. It is impossible for one person to do everything alone, so a team must be assembled to work together, but I question whether this can be achieved with a contingency fee model. People don't usually act unless money is secured up front.

Corporations also employ many capable personnel, and there are initiatives such as the Ministry of Internal Affairs and Communications'

"Regional Revitalization Entrepreneurs." I think it would be interesting if a successful model could be built. For example, using social-impact bonds to get a return commensurate with the outcome.
Abe: Yes. The increase in connected minds will bring in money from outside the region. It would be nice if municipalities would appreciate that impact.
Murakami: It would be helpful if social-impact finance could commit to that part. A corporate version of the Hometown Tax Program has been set in motion, but we need to be able to pull in more social-impact money on par with foreign countries.
Furuta: Quite a few people are willing to work hard for free at first. But the truth is, we need a cycle in which these people receive fees, produce results, and are duly evaluated, so the next person appears. A cycle of people who take risks and jump in feet first. The return does not have to be money. People who jump in feet first are not the kind of people who are motivated by money. That said, I think it will be really important to have something like a regional model of success that properly evaluates people who start from scratch.

Attracting a reserve army of connected minds in the form of "spectators"

Takemoto: There is always a possibility that accelerators will not be accepted by the region. No matter how competent, knowledgeable and well-connected that person may be. Accelerators who can bring a lot of the "spectators" mentioned earlier are the kind of accelerators who are most likely to be accepted. I really like the idea of calling them "spectators" rather than "tourists."
Abe: Just bringing people to an area is not much different from bringing tourists. They will spend money in the usual places, such as accommodation facilities. In our case, we are shifting the motivation for involvement in a region to the viewpoint of the payment side, such as transforming travel into learning during school trips, and transforming business development into human-resource development during corporate training. We should be able to broaden the supporter base organically with their desire to know more about and support the region. But we have not yet been able to properly monetize that aspect. That is why we have narrowed the scope of our activities, so we can charge for putting them

in the context of children's education or enhancing expertise in business development.

Takemoto: I see. So you are turning the situation around so as to be able to monetize spectators, and that is how you are able to make your business pay as an accelerator. However, if the people in the region were to do that on their own, it would be difficult, wouldn't it?

Furuta: Indeed, it may be difficult unless you are on the outside. In Mitoyo, for example, I am on the outside; Mr. Harada, who runs Udon House, is committed as an area organizer close to the region; and Mr. Imagawa, who runs Soichiro Coffee, is working on the inside. It is, in fact, thanks to this mille-feuille structure that it is possible to seamlessly carry out a range of activities both inside and outside the region. Sometimes I find it easier to bring someone in precisely because I am on the outside.

Takemoto: In the case of the Echigo-Tsumari Art Triennale, as the project developed, you could say that Fram shifted his position to the outside.

Furuta: By the way, wouldn't using the kanji character for "joy" be better than using the character for "spectate" in the Japanese reading of spectator, "*kankyaku*"? By which I mean those people will be happy to be involved in the region, or make the locals happy.

Takemoto: I think what I'm alluding to here is that it's fine to start as a spectator who just comes to see what is happening, and it will be even better if we can make some of those people happy.

Abe: I think there are various kinds of spectators. Some come just to take a look, while others want to make the area their stage. For example, supposing there are 100 people in a theater, and 10 of them watched the actor's performance and felt the urge to give acting a try. But the other 90 don't feel that urge, and are happy to continue to pay to watch. Such a situation is meaningful enough, in my opinion. However, I think it is important to design activities so that the people watching can sometimes changes places with the people being watched.

A profit-generating structure and digital platforms

Murakami: Whether or not there is a structure in place to increase profits after an initiative spreads in the area is an important question. Once that mechanism is apparent, money will pour in for sure. Reforming regional service industries is ultimately about changing the situation from one in which money is shared only within the local community to one in which new demand is uncovered and money can be drawn in from outside.

There are two potential patterns for making a profit. The first relates to trade logic, in which exchanging assets of different value is a source of profit. As long as people with the same values are exchanging equivalent assets, there is no way to make a profit. The question is how to bring into the regional economy a structure like that in which the price of rice and the price of wheat start to rise only when a country that only grows rice trades with a country that only eats wheat.

In other words, this corresponds to a structure that brings in spectators from the outside who are unaware of the value of the region. Working backward from the goal of creating a reproducible template for rural revitalization, one way to accomplish it is to accelerate the creation of such a structure.

The other pattern, which relates to the basic infrastructure, is the ultimate consolidation of service industries. By at least consolidating the services provided, the cost of each service can be reduced, services can be provided more efficiently, and the structure can be changed to one that is profitable.

After careful calculations, and hypothesizing which pattern will generate profits, those involved will reflect and decide on the key business, invest human resources, and create a reproducible model. I think it would be good for accelerators who join from outside and GPs who work on the inside to work together to create such a model.

Fujisawa: Listening to today's discussions, I feel that the most difficult component is the selection of projects. At that time, what is needed is a player to take on the role of Ridilover, a player who can work like Mr. Takashima, a player who can set up a business and a player who can talk about the future. I think this could be likened to a sports club, like a soccer or basketball club. For example, the J. League is trying to establish clubs in all prefectures in order to energize local communities, and they have even established a J. League 100 Year Vision. When a team plays an away game, many fans will follow. In a sense, they are the "spectators" in the Echigo-Tsumari example.

So if there is a region that cannot choose a key business, they should consider sports clubs as one example.

Abe: You either need a higher profit margin or a larger flow of people. In the case of Echigo-Tsumari, the buzz surrounding the Instagrammable tourist spot Kiyotsukyo Gorge (famous for the "tunnel of light") brought in many tourists, so we were able to raise the admission fee a little and invest that money in the NPO. The flow of people has increased and a profitable structure has been established. So if we utilize digital platforms in this kind of project, we may be able to create a smoother mechanism.

Session 7: Mr. Daisuke Maki's Case —Nishiawakura, Okayama Prefecture

Moving from population growth to income growth: Choosing not to merge and the fruits of the "Hundred-Year Forest Vision"

Nishiawakura is a small village with a population of only 1,400. Twenty years ago, after resolving not to undergo a municipal merger, the village risked its survival by initiating its "Hundred-Year Forest Vision" to revive the forests that cover 90 percent of its land as an industry. Now, after attracting new residents and launching a variety of businesses through the creation of more than 50 business ventures, it is exploring ways to expand employment and income. In this session, we speak with Daisuke Maki of the A-Zero Group, Inc., and other trailblazers about how this may be achieved.

Cultivating local ventures like forests

Maki: In addition to the project we will discuss today, which is located in Nishiawakura, Okayama Prefecture, the A-Zero Group has projects in the city of Takashima in Shiga Prefecture; the town of Atsuma in Hokkaido; and the town of Kinko in Kagoshima Prefecture. We also plan to expand our operations to cover eight locations in the near future. Under our mission of "contributing to future *satoyama* (land between foothills and plains that is used for agriculture and forestry) through projects in the domains of natural, social, and economic capital," we are also engaged in businesses including wood processing distribution, local venture development, and the Hometown Tax Program.

The A-zero (A0) soil layer is the top layer of sedimentary organic material in a forest; it consists of matter such as leaf mold. It is the A0 layer that protects and nurtures the rich layer of soil below, which is called the A layer. We named our company after the A0 layer to reflect our dream of restructuring and recirculating the economies of regional areas while respecting the nature and culture passed down among their communities.

While nurturing the seeds of various local ventures, our business utilizes grants for promoting rural revitalization and Local Vitalization Cooperators to drive and expand projects. This positive cycle can be likened to the growth of seeds into trees, which then become forests. We support the creation of new products and services from such projects. Then, with the increase of such attractive local ventures, donations through the Hometown Tax Program as well as the lineup of thank-you gifts increase. Since the A-Zero Group operates businesses related to the Hometown Tax Program and regional trade, our revenue grows when the amount of tax paid and the number of products increase.

In Atsuma, one of poultry farms has built a strong support base and is now attracting donations of around 100 million yen a year. Just as the A0 layer of soil helps to enrich the forests, we are able to enrich local communities by fostering many local entrepreneurs, which in turn brings more customers in a virtuous cycle.

We are also working to provide experiences. As various companies grow, marketable content also accumulates in the region. We then organize this content into experience-based services. One example is corporate training. We offer a forest experience program that deepens understanding of biodiversity and encourages participants to consider how to apply this understanding to corporate management.

The choice not to merge and a hundred-year plan

Hori: So what sorts of projects have Mr. Maki and his team been working on in Nishiawakura? We visited the village for this session, and here we will touch on how his work began.

A small village of around 1,400 people, Nishiawakura is located in the northeast of Okayama Prefecture at the headwaters of a river bordering Hyogo and Tottori Prefectures. It started a new journey in 2004 when it resisted a Heisei-era government initiative for strengthening administrative and fiscal capacity through major municipal mergers.

Instead of merging, the village bet its survival on a strategy aiming to revitalize the forests comprising 95 percent of its land as an industry. It resolved to honor the spirit of predecessors who had planted trees for future generations some 50 years ago by continuing efforts toward growing a beautiful forest over the next 50 years. It was from this aspiration that the "Hundred-Year Forest Vision" was born in 2008.

Ninety-five percent of the land in the small village of Nishiawakura is covered by forest.

The project is split into upstream and downstream operations. Upstream, a forestry cooperative co-manages privately owned forests totaling 3,000 hectares. Forest thinning and road maintenance are performed in a systematic manner to rejuvenate the forests and supply harvested logs to the village.

The downstream operations then work to develop products and businesses that utilize these logs. The cut timber is classified into three grades. Grades A and B are processed for use as residential building materials and are distributed as products nationwide, while grade C wood is used to supply electric heat to facilities in the village.

This ensures forest resources are circulated within the village and has led to the creation of a diverse range of businesses. The vision was launched around 10 years ago. Including sole proprietors, we now have over 50 business ventures, with a combined revenue of around two billion yen.

Maki: While the residents expressed their wish not to merge with neighboring villages through a referendum, the final decision was made by the then mayor, Masatoshi Michiue. Mayor Michiue had a negative view of the "distribution culture" imposed on regional areas and disliked the fact that funds received from the government would be divided with the merger. Instead, he wanted to focus on investment. Hideki Aoki, the current mayor, shares the same perspective.

When the Hundred-Year Forest Vision began amid these circumstances, I was an executive at Tobimushi Inc., a forestry consulting company owned by Yoshiki Takemoto, and I would visit the village with employees to coordinate work and build relationships. The 3,000 hectares of forests were owned by 1,300 people, so we worked steadily and negotiated with each of them. One by one, we got them to agree to a plan in which their land would be temporarily managed by the municipal office and used as a community resource.

The wood processing and distribution business in the downstream operations began with the launch of the Nishiawakura Forest School (*Mori no Gakko*), the predecessor company of A-Zero. It was jointly funded through the efforts of Tobimushi and the municipal office, with the external funds gathered via a crowdfunding model, which was quite novel at the time. That was back in 2009.

Children and workers increasing even as the overall population drops

There was a period when the Forest School struggled and ran at a large loss. However, we were eventually able to turn a profit, and we realized that if we could make such a recovery even in a declining industry like forestry, there must be other possibilities for success as well. In that context, we decided to launch the Nishiawakura Local Venture School to support new businesses. Our aim was to follow the "Swimmy strategy," where small businesses come together and benefit each other as whole. We tried to create lots of small companies and just have a go at various projects. This relates to the "warm-up stage" that Mr. Murakami has spoken about.

The municipal office also contributed by sending out a message with the catchphrase "Enjoy life" and a photo of the cedar forest filled with fireflies. Just as fireflies can light up the forest, the combined energy of individuals who brighten their own lives can revitalize a village.

People from outside now account for around 220 of the 1,400 villagers, and the number of children is increasing. Looking at the student numbers covering kindergarten, elementary school, and junior high school, they continued to decline until 2011. However, a few years after the Hundred-Year Forest Vision was launched in 2008, numbers began to increase.

According to the national census, the total village population is currently 17 percent smaller than it was in 2005. However, the population of children under 15 is not declining. There were fewer than 100 children under 15 in 2005, but now there are 80 more than expected. In other words, although the overall population is declining, the composition of the population is improving.

Looking at the neighboring municipalities, in the areas that merged into the city of Mimasaka, populations have fallen further below projected estimates. The number of children has also decreased in these areas, leading some schools to close.

Nishiawakura, meanwhile, has also seen very positive figures on the economic front. Disposable income, the number of taxpayers, and the average taxable income are all improving. I think this is because the working population is increasing despite the decline in the overall population. A similar trend has emerged in the town of Ama, as recounted by Mr. Takemoto.

Full-scale business needed for the next stage

Hori: Listening to this story, it seems the decision not to merge in 2004 played a major role in Nishiawakura's development. Akihiro Aoki, a local forestry business owner who we interviewed, mentioned that the current mayor once said, "If we try to do something and it doesn't work out, it's okay if we merge, but simply doing nothing and merging is not an option." His words still resonate with Mr. Aoki today.

While Mr. Aoki has lived in the village for generations, Sunao Tabata, who moved to the village from Tokyo five years ago, told us, "If you want to live a richer life, you have to think about a lot of different things." He launched a new company called Hyakumori after taking over a local forest-management business. Like Mr. Aoki and Mr. Tabata, talented people are clearly being attracted to the upstream operations.

We were also surprised when Mr. Maki, who runs a downstream company, told us that his work covers not only wood processing but also eel farming and deer-meat processing. When he decided to create his Mori no Unagi eel-farming business, he sent employees to a technical school in Chiba Prefecture for a little over six months so that each of them could acquire skills for individually handling all aspects of the business, from eel breeding to the *kabayaki* method of preparation and grilling. Thanks to the development of this new enterprise, village industry statistics documents now include a category for marine products.

Maki: That's right. In fact, one of our female employees who works in deer-meat processing also works in our nursing-care office. I think the whole region needs to think about how to put the right people in the right place. There are a variety of companies and jobs. We want to create choices and expand the potential of each individual.

Hori: So the aim is to fully harness the limited working population. This approach is being taken at the Forest School. While normally production lines would be combined for efficiency, at the Forest School, each employee is allocated their own workstation. This means that individuals can work comfortably at their own pace, doesn't it?

Maki: The line is designed to suit the needs of all employees, including those who cannot work from nine to five, such as women raising children. It enables us to have a lot of young women in our workforce, and the line can still continue if people are absent.

Hori: Nevertheless, although Mr. Maki and his team have been working diligently to build an industrial foundation in Nishiawakura, a dilemma has also begun to emerge. Takahiro Ueyama, who serves as the village's special counselor for rural revitalization, told us the following:

> A lot of local ventures have launched, and young people are moving into the area. While I think the quantity of ventures represents one key performance indicator, it's still difficult to reach the point where the employment of local residents increases. To reach the next step, rather than simply focusing on the number of companies, we need a model that can expand business scale and ensure strong capital.

To achieve this in Nishiawakura, we are trying to establish tourism as a pillar of industry. We plan to redevelop the ruins of the former lodge, which used to be run by the local government at a loss, and develop new tourism resources in collaboration with major capital providers. We are also looking into the development of digital platforms for achieving this. In our upstream operations, Mr. Tabata's company has improved the communication environment in the forests and released data on the forest environment to generate new businesses.

Maki: There is a strong feeling of excitement now that the warm-up stage has ended and a number of small-scale companies have been established. For the next step, someone must thoroughly design and create businesses. We are now exploring how to find general partners (GP) that can take on this challenge and any risks involved.

Protective measures to keep ownership in the community

Hori: When we saw Mr. Maki's operations firsthand, it was clear that everything was designed to suit individual needs and was the result of careful and consistent communication. The wood-processing line was eye opening. If you were moving into the village, these sorts of unique efforts toward securing income opportunities would be very reassuring.

Miyase: Nishiawakura is a place where people moving into the area—so-called I-turn residents—can truly put down roots. I also believe that more employment opportunities will emerge for young local residents and those who have returned to the area, or "U-turn" residents.

Fujisawa: I was impressed. I'm moved by the young staff members who worked hard to acquire skills in eel processing and deer butchering. Those in forestry also seemed to take great pride in the community. In the space of 10 years, the villagers have truly changed.

Hori: As someone in the forestry business, Mr. Aoki told us: "We are now much more aware of the importance of adopting new ideas. That is the path we must take to protect the village. For Nishiawakura to survive, we need to always be a step or two ahead of the rest."

Takemoto: In the beginning, I went to Nishiawakura with Mr. Maki to help on projects. I met various different people then, including an elementary school student who now works at the Forest School. It's very moving to see how they have grown.

It's also great that the local ventures are sharing human resources. In Ama, we have a system called the Multiple Work Cooperative that rotates newcomers to the town across multiple workplaces. However, the system in Nishiawakura is different; it's remarkable because it operates

more flexibly on a private-sector level.
Furuta: Like Mr. Maki said, it's about how to build the next stage. The situation in our city, Mitoyo, is similar. While it's great that we have lots of small lodgings, they are not sufficient to provide accommodation for the numbers of people now visiting the area. From that perspective, it might be worth building a large hotel with outside capital. It's a question of how far we can go.

So far, we have worked together as a single team, but the next phase will focus on increasing business scale. It will be a transition period, and I think an important theme will be how to maintain local ownership. It's okay if the capital comes from outside. However, it would not be good if ownership were also outside the community, as this could stop employment from flowing to local companies.

On the Avenue des Champs-Élysées in Paris, many buildings are actually owned by wealthy people from places such as Dubai and Doha. However, the government prohibits changing the form of the buildings without permission. In other words, the rules of France must be complied with, and this ensures that a sense of ownership still belongs to the community. The buyers accept these conditions and hope to become a member of the area while respecting French culture. This works like preferred stock, which is without voting rights. We are currently thinking of building this sort of framework to create unique relationships with our region.
Maki: That's right. The important thing is focusing on where we go from here. The number of regions working on similar initiatives is increasing, and the market is becoming more competitive—a red ocean, as they say. We are reaching a stage where we can bring in outside companies that are well capitalized and truly want to collaborate with the community.

Building systems to attract external funding

Furuta: You mentioned earlier that local ventures are increasing, but actual employment is not rising significantly. In the first place, is it right to expect startup ventures to create and expand employment?
Maki: Even if ventures do not create much employment, there is an understanding among locals that they are still good if they are working enthusiastically. However, what we really hope to see is companies that make people who have left the village want to return, or that enable

Fig. 1-10. Takibi program

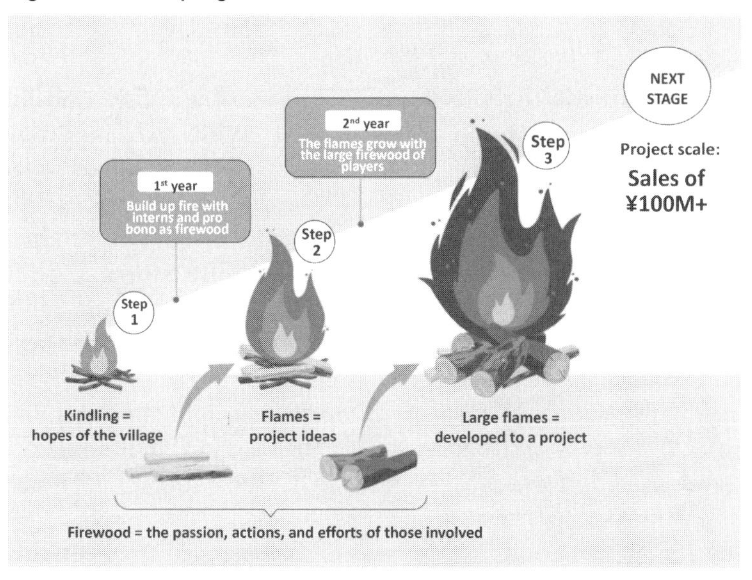

school graduates to find employment in the area.

In light of this, we launched the Takibi ("bonfire") program in spring 2021. As a successor to the Local Venture School, this program has aimed to create new businesses while having a system in place to attract outside companies with strong capital.

While we have fostered entrepreneurs in the village after bringing in ideas from outside up to this point, we are now aiming to develop concepts locally and involve various people from here. Business concepts should align with the wishes of local residents. For tourism ideas, for example, we could bring people together and discuss things over two-night, three-day stays. In the initial stage, we will focus on local participants, and then gradually increase the participation of outside parties who may wish to work with us. In this way, we will expand relationships and ideas little by little, like building a bonfire, with the aim of nurturing businesses that can accommodate U-turn residents.

Hori: If you expand tourism and outside companies actually do come into the village, couldn't that fragment relations with the village?

Maki: In tourism, I think it will be hard to generate value if we don't collaborate with external ventures. I also believe that Nishiawakura's forest-oriented manufacturing base is seen as an advantage by outside companies.

When looking at the village's GDP, it is clear that the contribution from manufacturing is rapidly increasing. With the Hundred-Year Forest Vision as the foundation, we now have a forest-linked production system encompassing felling and bucking, wood processing and interior materials, and sales. However, if we focus only on production, it will be hard to generate further value in the future. How can we create business that will take us to the next level? I think we can expand from manufacturing to services, stepping it up to the development of services linked to the use of local resources.

For this, digital platforms such as forest geographic information systems (GIS) can play an important role. As Mr. Hori mentioned earlier, the development of practical systems that utilize forest GIS is already underway in Nishiawakura.

The important point is that we are making steady efforts toward moving from a primary industry to a tertiary industry. Rather than simply bringing tourism into the area, we are striving to position new businesses on a sound, community-built foundation. I feel that this puts the community in a stronger position and helps to preserve its identity. Without taking the time to create this layer, it is hard to suddenly come up with services.

Furuta: I fully agree. It's about having a business foundation within the community and then adding elements from outside on top. With this sort of model, community ownership can be preserved. As is often the case with tourism, when land is sold and a hotel is built, the services are usually handled internally, and the local community is not involved at all. Your approach is completely different.

Maki: I think the Takibi program is an excellent example of a framework that was carefully designed for this purpose. When companies come to Nishiawakura through the program, it is critical to engage in thorough discussions. While the municipal office provides appropriate subsidies, ventures from outside the village also commit to an appropriate amount of self-funding. The challenge is creating as much synergy as possible through the capital on both sides.

Retaining GP candidates who take on challenges

Hori: For this discussion, Maki asked everyone to do some homework by thinking about the following topics: methods for translating ecosystems into content; how to create hubs in the region for attracting visitors; building interesting frameworks for local ventures; capital alliances with major companies, and paths that can be taken for the future. Mr. Furuta, would you like to share your views?

Furuta: In Mitoyo, we created a dormitory-like facility called Gate, recognizing that it was essential to have a place where those wanting to take on challenges in the region could stay. It serves as a hub for people within and outside the region to connect, and encourages migration to the region. Around 300 new residents moved to Mitoyo last year.

So instead of suddenly starting a typical accommodation business, we created a facility like a dorm to meet their needs. Rather than serving as a site for attracting tourists, it aims to bring together people who can link the area to tourism. I think it is important to have this type of facility in place.

Hori: The basic aim is to attract GPs and funding for the rollout stage of rural revitalization and ensure that efforts contribute to improving the income of villagers.

Furuta: A bottleneck in this process is the significant lack of human resources. This is a challenge we are particularly struggling with. That makes it even more important to create places for attracting a diverse range of participants. These could be hubs for gathering local ventures under the Swimmy strategy, or places akin to *ikesu* ("fish tanks"), where people wanting to get involved in the region can grow into workers. I believe these places in themselves can become marketable content.

Maki: It would be great to have room to invest in that idea. However, it can be a challenge to get the municipal office to commit. The fact is that it is difficult to invest in things that are hard to understand, even if they might be important.

Furuta: If the village already has 50 local ventures, how about each company puts in 500,000 yen? That way, I think a small dormitory could be created by renovating a private home. This type of place is important, as it gives newcomers a chance to test out the village, and once they become comfortable, they can move to a new home. To be honest, I also had my doubts before we built Gate, but once it kicked off, plenty of people came to stay. I think this sort of mechanism is necessary, as it acts as a cushion for new residents.

Attracting investment through social impact

Miyase: Ms. Fujisawa, what do you think about the possibility of attracting large-scale investment through collaboration with major companies?

Fujisawa: Until a few years ago, it would have been difficult, but I think things have changed now. One reason for this is the emergence of social-impact investing throughout the world. We have thoroughly evaluated the social impact of the A-Zero Group's business, and we clearly explain how companies that invest in us contribute to this impact. Contributions do not need to be purely monetary; people and technology are also helpful. If we explain how the A-Zero Group uses such contributions to create social impact, it becomes easier for investors to include us on their balance sheet, and it can also help their stock value. I feel this is one attractive point.

Another advantage of Nishiawakura is its forest. The A-Zero Group clearly states the carbon credits that our business can earn, and companies that invest in us are able to obtain these credits. This is a positive aspect for investors, and it can help to spark collaboration.

While this is a completely different form of return on investment from conventional returns, if this sort of framework is carefully designed, it becomes easier for global companies and family-owned companies to get involved. I feel that family companies, which are often run with an eye to the next 50 years, have a strong attachment to their founding areas and want to contribute to their communities. When such companies invest in the A-Zero Group, they can gain knowledge that is helpful for revitalizing such areas, and this in itself becomes an incentive for investment.

One other method for attracting investment is issuing nonvoting shares or other completely new types of stock certificates.

Takemoto: Our town has also had a lot of discussions about corporate investment in relation to the forest, and on topics like stock certificates without voting rights. There has also been talk about creating a fund from large-scale corporate investments and then using the yields from overseas assets to invest in the village.

For over 10 years now, Mr. Maki has been exploring ways to shift from a modern capitalism model to a new approach that can meet unique needs. I think there is an increasing possibility that the A-Zero Group could realize this goal.

Methods for leveraging the forest as a powerful asset

Hori: Mr. Kamiyama, what do you think? For example, would it be possible to adopt an *albergo diffuso* (decentralized hotels) approach?

Kamiyama: Yes, it's possible. However, I think an *ospitalità diffusa* (scattered hospitality) approach would work better here. The focus should be on creating a hotel business operating across a wider area of the region.

Nishiawakura's asset is its forests. Yet, at the moment, it is not being used for purposes other than timber. What if we apply the *ospitalità diffusa* concept to the entire area, including forests, and gradually expand a business for treehouse-like accommodation facilities? I think it would be fine to start with just a few buildings. An owner should also be decided. This could be Mr. Maki's company or one belonging to another person. Then, once it reaches the point where the owner is seeing returns, the project could be expanded through selling lots in a few areas. This could include traditional village homes (*kominka*) as well as forest land—literally the whole region.

Visitors could experience the forest and mountains and get to feel like a real resident during their stay. I think there would be ways to launch projects with this concept. Why don't you get me involved and we can work on things together from the fund-building stage?

Maki: Now I'm feeling even more motivated! [Laughs]

Hori: I think study tours could also be run with Mr. Abe.

Abe: There is definitely a strong possibility. However, to turn the focus to something more immediate, what I have noticed through today's discussion is that there is intense competition among a very limited number of players in the region. I think one challenge is how to secure committed entrepreneurs who are willing to work in this very competitive red-ocean environment. I'm talking about special players who can generate revenue in the hundreds of millions of yen per year.

In our case, in addition to keeping ownership in the community, one of Echigo-Tsumari's policies is to find special players and cater to their unique needs. For this, we need sharp judgement when choosing players. I think this sort of strategic approach is also essential for Nishiawakura.

Maki: Thank you, everyone, for thinking about this in such detail. You're certainly right about impact investing and carbon credits. In our village, I think we will handle role allocation through the Hyakumori company, which also takes care of forest management. My company will oversee biodiversity. The challenge will lie in how we apply indicators to these initiatives to measure their social impact.

Fujisawa: Mr. Maki, I'm sure this area is your forte.

Attracting people to the "kelp forest" in order to ensure an adequate workforce

Murakami: Building a model for successful rural revitalization can be divided into three steps: the warm-up stage, the rollout stage, and the expansion stage. Today, we have heard a lot about how to transition from the rollout stage to the expansion stage.

It seems that just before entering the rollout stage, it is essential to create systems for providing human resources to the region in a structured manner. Without such systems, even if we have good ideas, subsidies, and funds, there can often be a shortage of people who can reliably manage businesses once they are established.

Furuta: That is an excellent point. In Mitoyo, we were lucky to have local residents who made great managers. However, if they had not been there, such shortages could have caused a bottleneck that kept us from advancing to the investment stage. In Mr. Maki's case, some people entered the warm-up stage through the Local Venture School. The challenge is the next step, though.

Hori: Do these types of workers usually come from within the community? I would think that some regions may be lacking such people.

Furuta: I believe that each region definitely has potential human resources, but the issue is how to create an environment that encourages such people to get involved.

Abe: I agree. Sometimes the key people in the businesses can actually reduce opportunities for involvement. There is also no real business potential if things are stuck at a stage where people are just spectators. Unless these obstacles are eliminated, it is impossible to move forward.

Kado: There are people, at least in the beginning. So it's a matter of having a "magnet" in place that can draw these spectators to the next stage. If we don't do this from the start, it's hard to attract people, even if we try to later.
Furuta: You might want to suddenly do business together, but it's not always easy. That's why we started with a trip to Poland to observe examples.
Abe: Yes, in the beginning, I think it's good to have something that can motivate people by sparking their ambition. Then a different incentive is needed to move them to the next stage of becoming real team members.
Furuta: That's right. It could be something akin to a corporate training system that both provides training and encourages participants to stay afterwards. We are currently running two or three observation tours a week, and if participants mention that they want to come back, we ask them if they are interested in staying a week. Usually one participant per tour will actually then go on to stay at Gate. So rather than just creating good spectators, I think there are multiple ways of attracting people to the "kelp forest."

Communicating stories from the community with intention

Murakami: I think that communication gradually helps us to reach this goal, and it should be used strategically from quite an early stage. If you leave communication until you actually need people, it's too late. The time lag must be kept to a minimum.
Miyase: Mr. Maki's company has been communicating through A-Zero's owned media. It uses a platform called "Through Me" to present people and events in the community.
Maki: While we are not at all focused on things like page views, quite a few people get in touch with us after seeing our articles. As time goes by, a great article can hook audiences, so it's important to archive them properly. Our article "Yamada-san no Hitsuji" (Mr. Yamada's Sheep), for example, is long-tail content that has always been popular with readers.

In the warm-up stage we experience various challenges, and we use media to carefully record random events so that we can see how they connect to the future. Whenever there is a story about someone's challenge, we interview that person and communicate their story. We have been making a conscious effort to do this, as it helps to link our work to the community's vision. It's really about being able to make the most of whatever may occur.

Miyase: Nishiawakura became well known partly because key phrases like "the village that didn't merge" drew a lot of attention. How do you think other areas without that background can communicate information and attract interest?

Abe: Each region needs its own symbolic stories, and it's not possible to bring them from outside. For example, at the Echigo-Tsumari Art Triennale, we placed a series of works called *Kabakov's Dreams* at the first entrance. These were created by Ilya and Emilia Kabakov, artists from the former Soviet Union. Their works depicting farmers in harsh and snowy conditions conveyed the region's story.

While regional symbols like Instagrammable attractions are important, it's hard to communicate what makes a region special if local stories do not come first. A region does not need to be reported on by a major media outlet; rather, communication can start simply by having local people talk to visitors. Things should be designed in a way that then enables media providers and social-media influencers to expand communication from there.

Steps for raising productivity and increasing earnings

Murakami: As the flow of people into the region becomes more substantial and key businesses continue, a combination of leaders and coordinators for the next set of businesses should be put together, evoking a mille-feuille structure. That's the rollout stage. But what about the final stage, expansion? While key businesses help in shifting from a distribution culture to an investment culture, this alone is not enough to change a local economy. The final goal is to increase the income of local residents and to properly build highly productive service industries to achieve this.

How exactly do we create highly productive businesses? I would like to raise three points based on today's discussion.

The first point is assembling key businesses while investing in digital platforms at the same time. For example, this could be a model that combines the My Number Card system with efforts to attract tourists, which I spoke about in the previous discussion. This could potentially speed up the process of involving surrounding parties in key businesses and minimize the need for extra investment. I believe that using digital platforms to raise productivity means increasing cost efficiency and supply volume in local service industries. This can be thought of as one model.

If the ripple effect of such efforts leads to the launch of associated service industries, it becomes easier to build a structure for increasing wages and employment. This can be likened to what Mr. Maki mentioned earlier, where a foundation can be created by building up manufacturing, and then service industries can be established on top.

The second point is attracting capital by skillfully leveraging the community's strongest assets. For Nishiawakura, this is the forests. A digital platform would also be helpful here if the village hopes to attract capital through carbon credits. A forest GIS is a good example. This type of platform is important because credits can only be applied to things that can be measured.

With a digital infrastructure in place, the scale of capital can be expanded dramatically once a certain stage is reached. In turn, the advantage of scale will help to increase productivity and income for each local resident.

The third point is the incorporation of trade logic, which I also spoke about last time. A basic principle of economics is the creation of profit through the exchange of different forms of value. Productivity cannot rise if this exchange does not happen.

In this sense, I think that Nishiawakura's strategy of strengthening tourism and inbound demand is the correct choice. With continually declining birthrates, consumption markets across the whole country, including the Tokyo area, will contract. Relying solely on domestic product sales to generate profits will no longer be a viable strategy.

In light of this, more companies are likely to aim for profits through charging high fees on high value-added services, or by selling products overseas, or having people visit Japan. So how can a framework for this be assembled in an environment consisting of key businesses and a digital platform? I think this type of economic strategy will be put to the test.

Roles and positioning of the public and private sector in regional management

Furuta: The positioning of the government is also quite important. It is quite difficult to expand businesses and build a digital platform when working solely as a private enterprise without any government connections. However, the government and the private sector are not always aligned in their perceptions, especially in terms of speed. I feel this can also be an issue.

Murakami: When I discussed this with Mr. Abe at the Ridifest (an event organized by Ridilover for discussing social issues), he mentioned that the government is skilled at distributing a limited pie, while the economy is skilled at distributing a growing pie. Previously, and especially during Japan's period of rapid economic growth, the entire market—that is, the full pie—experienced growth, so the government and economy were not in conflict. However, now that the domestic market is starting to shrink, we must once again examine what the government can solve, what the economy can solve, and how public-private partnerships can work in the future.

Abe: I think the government's next role can be to focus on identifying new challenges in the market. Challenges so far have centered on how to build material wealth and how this wealth can ultimately be redistributed. However, there are now too many issues in regional communities, and the government cannot cover everything through taxes alone. Therefore, all issues must be evaluated comparatively and ranked according to importance and priority. The administrative and political side should assess the cost savings and benefits obtained in case the market solves such challenges.

Rather than being based on something like a petition, this comparative evaluation must begin with consensus-building within the community. That way, correctly ranked issues are linked to market evaluations. The private sector will then have greater incentive to contribute because it will be involved in solving issues of a higher priority or importance. The role of the government should be to prioritize issues appropriately and create ways of encouraging private enterprises to contribute. This is how I feel it could work.

Furuta: In Mitoyo, we are now working on data linkage. For example, income from community bus fares is extremely low compared to the required budget. Yet despite the large loss, the budget cannot be reduced. For nursing-care facilities, costs are some 100 times greater. Since the likelihood that a senior citizen will require long-term care increases as their physical activity decreases, it is estimated that 35,000 yen in annual medical expenses can be saved for every person who is able to take an average of 1,500 steps per day. So when data like this is looked at as a whole, it reveals that it may be better to invest toward increasing community bus services to encourage the elderly to go out and be more active.

Establishing a digital platform makes it possible to identify these types of causal correlations. Social impact can then be measured more effectively, and if the government shows where investment should be made, the private sector can enter more easily. I feel it is now essential to develop this kind of mechanism.

Murakami: That's right. While the private sector may be able to move from the warm-up stage to the rollout stage alone, it is impossible to advance to the expansion stage without working with the government and sharing a sense of direction. Since data serves as an objective indicator, it is essential to create a venue where both sides can support it effectively. This in itself can help to attract investment.

For this, it will also be critical to have people who can analyze data, like local data scientists.

Furuta: We need people who can maintain a management perspective and monitor and reprioritize key performance indicators as necessary. Since this tends to be a weak point for the government, I think we will see more instances in the future where the private sector enters and reforms management styles. It will be a more corporate approach to regional management.

Building a highly productive local community

Kado: While data is of course important, it is vital to outline a framework that enables everyone to get along. The idea is for everyone to come together as a team and collaborate naturally.

Kamiyama: Municipal leaders play a key role here. Ultimately, rural revitalization is driven by municipalities, and therefore nothing will change unless their leaders give specific instructions. When it comes to doing something in a community after coming from outside, I think it is very important to look at how the municipal leader thinks and how much energy they commit to things.

Also, while it can be good to have groups from outside come in and expand in numbers to transform a community, even strong and sudden efforts can sometimes end up simply increasing conflict. It is crucial to gain the understanding of locals so that everyone can be on the same page. The same can be said for the municipal council. Even if the municipal leader wants change, some level of agreement must be gained from the opposition.

Hori: When meeting people in Nishiawakura, it's easy to get the impression that various professionals are brought in from outside. However, some of them are local residents. The workers in eel and deer-meat processing are a good example. In the civil society we live in, everyone is a professional at something, and the regions that are able to tap into this talent are strong. All of the rural revitalization project sites I have visited as part of this Trailblazers Conference belong to such regions.

Fujisawa: When you think of the word "productivity," what usually comes to mind is how to increase output without changing the amount of input. However, in light of what Mr. Hori has said, productivity can also be improved by harnessing the untapped potential that exists in communities. I think it is essential to explore this approach, and that doing so will also lead to changes in the ways digital platforms are used.

Murakami: I would like to add a little more before we finish. So municipalities put in hard work and the private sector gets involved, but what can the national government do? Firstly, it could be responsible for institutionalizing the organization of accelerators. The aim is to have GPs and area organizers inside the region and accelerators outside. To set up teams in this way, I think it's critical to have a system that ensures accelerators who can work over wider areas throughout the country are in place.

Secondly, the national government could oversee open-source development of digital platforms. I think it should provide this at a low cost; we could begin by seeing how well solutions using the My Number Card system work. This could be the first important test.

Finally, I would like to see the national government create opportunities for communities engaged in rural revitalization to collaborate horizontally. Just like we have noticed through the stories in today's discussion, everyone is facing similar challenges. To ensure our efforts do not become divided and isolated by area, I believe that the national government can create a space that can keep us connected.

Chapter 2

Rural Revitalization Trailblazing Model

We must never allow the ongoing examples of success in rural revitalization to simply end with, "Good job. Well done." This is why we want to extract elements from these examples that can be applied to any region and create a model for success. If we could engage in discussions with people who have succeeded and who can analyze examples from other regions, I have no doubt that we could begin to see what this model for success looks like.

This is how the Trailblazers Conference began. Based on the discussions we have held to date, we have divided this model for success into the three stages below.

(i) Prepare (warm-up stage)
To begin rural revitalization efforts, it is first important to improve the region's diversity. This does not mean you have to quickly start encouraging people to move to the area. What is needed is a gradual, gentle increase in interaction with various people from other regions. While this can begin from the rollout stage, it is also important in the next stage to work out how to incorporate elements from activities in the warm-up stage.

(ii) Identify a key business (rollout stage)
In this full implementation stage, various methods can be used to move forward and make progress. However, it is first important to identify a key business. Here we recommend introducing a digital platform to support such a business. It is also at this stage that we recommend utilizing the Type 2/3 grants for promoting the Vision for a Digital Garden City (digital implementation type).

The key is to set up a solid business based on the focused topic, and examine the extent to which you can involve local businesses and citizens. It must not conclude with one-off technological demonstrations, but be an ongoing business supported by locals.

Fig. 2-1. Rural revitalization trailblazing model

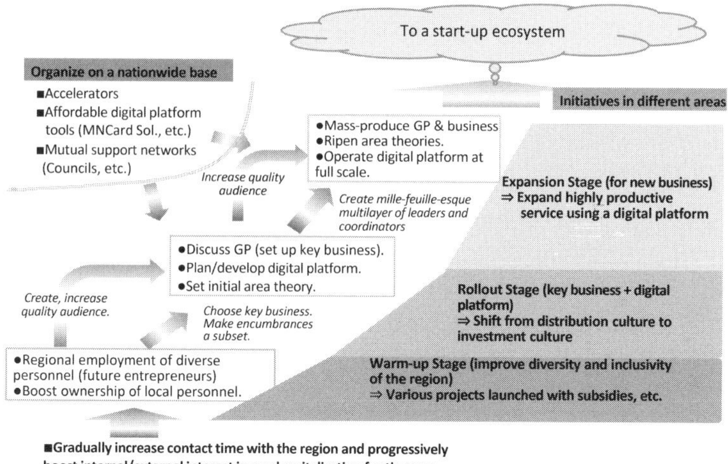

(iii) Consider key points ahead of expansion (expansion stage)
Once you the key business is established, the next step is to share a digital platform without delay while at the same time working on the launch of related businesses. While there are many different possible approaches, the aim is to create a start-up ecosystem. This involves soliciting the help of various external experts and successively setting up a range of different new businesses, including those that locals will be responsible for. The ultimate goal of these initiatives is to link the ecosystem to the market of financing.

To get these initiatives underway, you absolutely must have a strong sense of purpose. In this chapter, we examine what this sense of purpose looks like before delving further into the three stages described above.

1. Developing a strong sense of purpose

(1) Set a common goal across various business operators and citizens

In Japan, we often use the term "regional revitalization." However, what needs to be revitalized for it to be considered regional revitalization? Taking a step back to think does not make it any clearer. In this sense, it is indeed an ambiguous term. But why have we continued to use it, despite its ambiguity? Probably because the ambiguity was convenient for individuals from various fields and various standpoints to provide a justification for launching subsidized businesses.

Conversely, a clear definition gives little room for a person to use it as a justification to start their own business. However, there is no room for ambiguity once you enter the rollout stage of rural revitalization. If individuals were free to conduct a range of different initiatives without an overall framework, there would be endless inconsistencies and repeated demonstrations. As we note later, it would be like climbing a mountain without a summit.

If you are simply going to conduct whatever demonstration you choose in each different field, it may be convenient not to have a summit. But the only people who would be happy with this are those who have obtained a budget for implementing demonstrations of their businesses. This approach is unlikely to generate any results or changes that benefit local livelihoods. What we advocate is that, when you operate different businesses across multiple fields, they should aim for a single objective (summit) in a way that will really help change people's lives.

These efforts can start from any field. For example, trying to start a new type of social education with the help of local senior citizens could lead to the setting up of transportation to access this education. This transportation could then also be used give locals access to medical care in the area. It is important to have businesses work together in these kinds of ways for mutual benefit and to develop with close ties to the region. When looking at where to start, it is important that the businesses working together have a common vision and goal, or at the very least are aware of the importance of doing so.

Fig. 2-2. Climbing a mountain without a summit

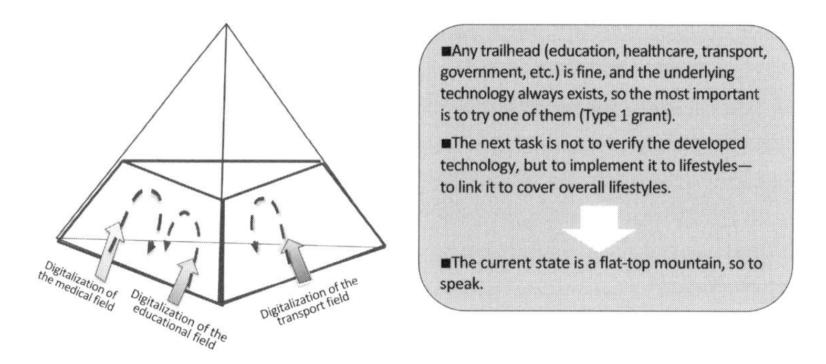

Fig. 2-3. Mountaineering is more enjoyable when you can see the summit

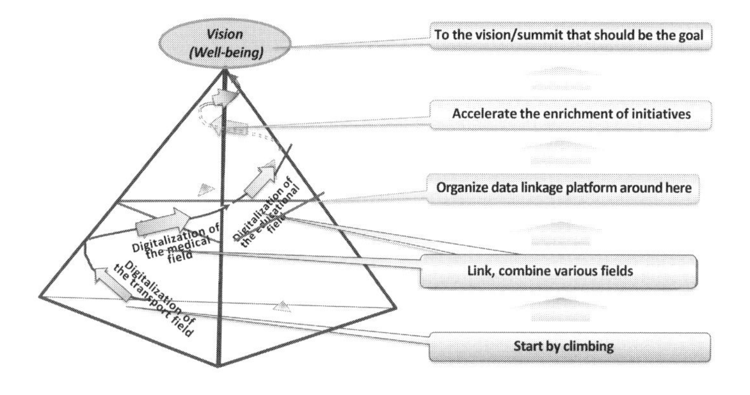

(2) Types of shared goals

What should this common goal look like? Let us take a look at a general example.

To date, the government has advocated rural revitalization and promoted a range of related measures. For rural revitalization, the biggest challenge has been the excessive concentration of Japan's population in metropolitan areas. The government's aim has been to stop the outflow of people to urban areas, encourage young people to gather in rural areas, and build a Japan for the next generation.

Fig. 2-4-a. Labor productivity trends and presumed causes

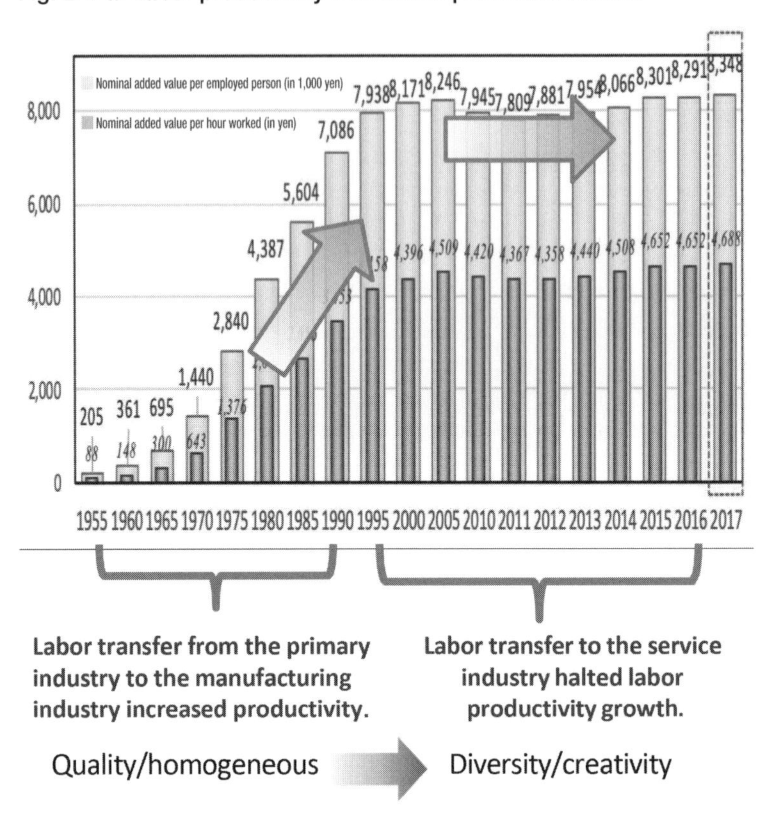

A typical example is snow shoveling and snow removal in snowy regions. If young people continue to leave rural areas and tasks like these are not fulfilled, the number of regions that are no longer livable will grow. In this sense, the help and support of young people is paramount. Unfortunately, though, the disparity in lifestyle quality and income between rural and urban areas is only likely to spur a rural exodus. It is therefore important to use the abundant natural and environmental assets in rural areas to create an attraction that is on par with the different appeal of urban areas. This will help to create regions where young people want to gather and lead modern lifestyles. The "Vision for a Digital Garden City Nation" that the government is currently promoting is intended to accelerate these activities using digital technologies.

Fig. 2-4-b. Output trends in the domestic manufacturing industry

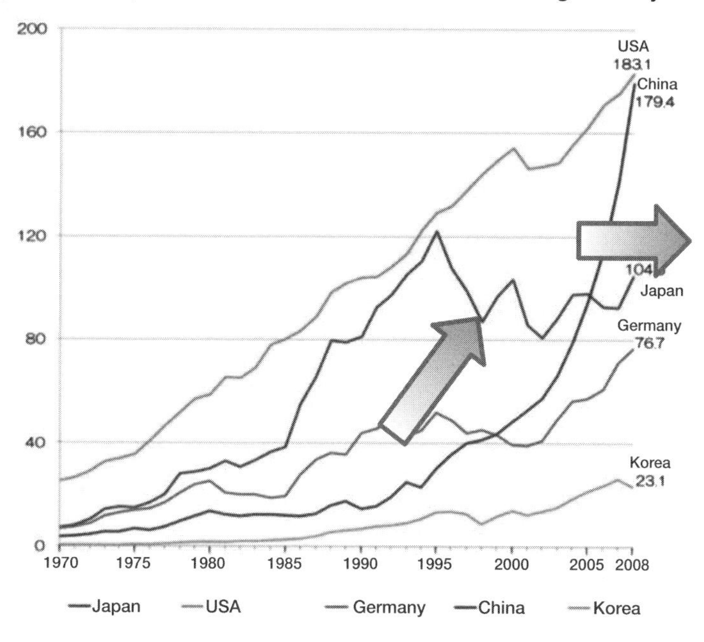

Source: National Accounts Statistics: Main Aggregates Database (United Nations Statistics Division), as of 2010.

It may seem obvious, but the availability of attractive jobs is essential for realizing the Vision for a Digital Garden City. Even if the region offers contact points for people interested in relocating, or a quality environment for raising children, without jobs to draw young people from urban areas, it will not be possible to develop people of the next generation who settle down and follow current rural lifestyles. An important point to note, however, is that many of the available jobs in these rural areas are low in labor productivity.

Let us take a look at why this is a problem. First, a general economic theory. Please bear with us, as it is an important point to understand.

Labor productivity is profit per worker, or more precisely, the value

Fig. 2-5. Productivity by industry

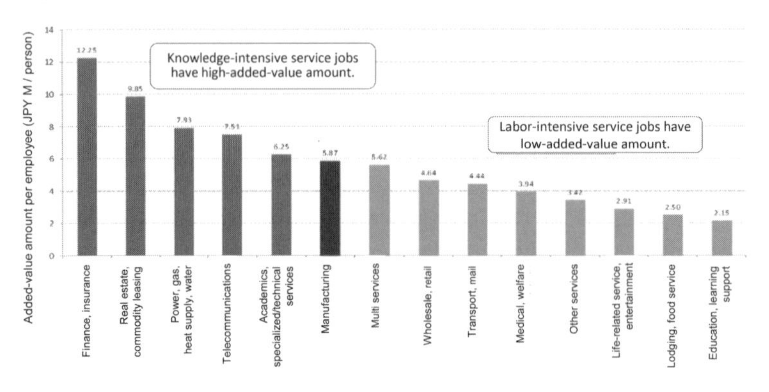

Materials: prepared with reference to Ministry of Internal Affairs and Communications "2012 Economic Census - Activity survey" and "2010 System of National Accounts"
Note: Added-value amount refers to that of the 2012 Economic Census.

added per worker. If labor productivity does not improve, employee wages do not increase. For example, even if sales increase threefold, if this means that three times the number of employees are required, wages cannot rise. To increase wages, the amount of profit per worker must improve.

Figure 2-4-a shows long-term trends in Japan's labor productivity. Since around the year 2000, labor productivity has hardly risen at all. As shown in figure 2-4-b, this period roughly correlates to the time when domestic production in the manufacturing industry stopped growing.

Even now, areas that depend heavily on manufacturing have maintained high levels of productivity and real wages. This does not apply only to the manufacturing industry. Without highly productive industries in a region, the wages in that region will not increase.

Now take a look at figure 2-5. The graph shows labor productivity by industry; manufacturing is right in the middle. To the left are urban service industries. And to the right are lifestyle-based service industries, which are common in rural areas.

What can we learn from this information?

During Japan's rapid economic growth period (from the mid-1950s to the early 1970s) in the Showa era, people moved from primary industries with low productivity to the rapidly growing, highly productive manufacturing industry. As the working population moved from a sector with low productivity to a sector with high productivity, both labor productivity and wage levels automatically increased in those regions and in turn across Japan.

However, as domestic production in the manufacturing industry stopped growing, and the industry's ability to employ additional personnel weakened, the only people who benefited from wage increases were those who found employment in urban service industries, which boasted higher levels of productivity than manufacturing. And then, large numbers of workers who were not taken on by the manufacturing industry ended up moving to service industries with lower levels of productivity than manufacturing. As a result, apart from those who left for the city and managed to find jobs in large companies, many found that neither overall productivity nor wage levels were increasing. This is the current state of the Japanese economy. It is particularly noticeable in regions where there is a high level of dependence on these types of service industries. And so, when thinking about restoring regional economies, it is first important to look at how to improve the productivity of service industries with low productivity.

(3) The importance of digital technologies

In this section we look at why digital technologies are essential for rural revitalization initiatives.

For service industries to increase profit, it is important to look at the density of demand in the area they serve. Whether it is tourism, accommodation, restaurants, or education, to boost profit with few service personnel, the most effective thing to do is to conduct business where there are lots of customers. That said, populations in rural areas are beginning to decline, and so for service industries there is only hardship ahead. This is particularly apparent in the change in the supply–demand relationship in terms of which side accommodates the other.

Figure 2-6 shows the change in industries where the demand side has aligned with supply-side circumstances. As populations decline, the supply side will have no choice but to proactively align with demand-side circumstances.

Let us discuss an example. Demand does not suddenly disappear when populations decline. But if we use transportation as an example, as bus or train drivers age and the number of working drivers decreases,

Fig. 2-6. The shift from the supply-side oriented economy to the demand-side oriented economy

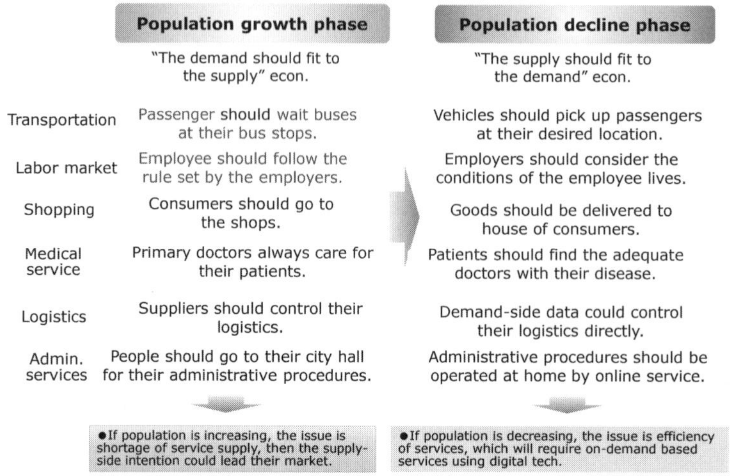

	Population growth phase	Population decline phase
	"The demand should fit to the supply" econ.	"The supply should fit to the demand" econ.
Transportation	Passenger should wait buses at their bus stops.	Vehicles should pick up passengers at their desired location.
Labor market	Employee should follow the rule set by the employers.	Employers should consider the conditions of the employee lives.
Shopping	Consumers should go to the shops.	Goods should be delivered to house of consumers.
Medical service	Primary doctors always care for their patients.	Patients should find the adequate doctors with their disease.
Logistics	Suppliers should control their logistics.	Demand-side data could control their logistics directly.
Admin. services	People should go to their city hall for their administrative procedures.	Administrative procedures should be operated at home by online service.
	•If population is increasing, the issue is shortage of service supply, then the supply-side intention could lead their market.	•If population is decreasing, the issue is efficiency of services, which will require on-demand based services using digital tech.

public transport at night becomes less convenient. For example, what should someone do when they arrive at the station closest to their home at 9 p.m.? There are no buses at night, and taxis are infrequent. In regions like this, senior citizens who have surrendered their driving licenses due to old age will have trouble getting to the hospital. Moreover, providing a diverse education could also become a challenge. For example, children who wish to take part in educational projects to observe ecosystems at the beach while picking up rubbish will find it difficult to get there with no transportation. This in turn will mean that the sphere of activity for people in these regions gets smaller and smaller.

This is where things get strange. Even though there is still unmet demand, local bus and taxi companies are struggling. This is the reality. There is a strange phenomenon where, even though there is excessive demand, people on the supply side continue to struggle. This no doubt sounds very different from what you learn in basic economics textbooks.

But in fact, it is population decline that holds the key to unraveling this phenomenon.

When populations rise, bus and train companies can increase the number of services or routes to meet the requests of the demand side.

But the problem now is that it is not possible to increase the number of buses or the number of drivers. In this situation, it is difficult for the supply side to use its limited resources to meet increasingly diverse needs. With a limited number of drivers or vehicles, teachers, doctors, or shop staff, it is difficult to cater to scattered, low-density demand. There is no choice but to accurately gauge demand-side trends and effectively deploy the limited resources on the supply side. A digital platform that can share demand-side data with the supply side is essential. This is because there is no other way to meet demand and increase productivity on the supply side.

Alongside these essential digital platforms, we will also need mutual-aid models. What do we mean when we talk about digital platforms and mutual aid? Let us use an example of a digital platform that can be used to obtain demand-side data. In the future, it will be important to be able to access real-time data on transport demand. But even if individual companies such as local taxi or bus companies were to invest in these systems, no one would see a return on their investment. And so no one invests. As a result, it is almost impossible to provide digitally driven services.

We are now seeing the collapse of both mutual aid and independent financing. And regions in Japan are completely stuck in this vicious cycle. They cannot create services that they know would be successful. This is the very meaning of the digital defeat.

When populations were rising, public aid only had to be provided to infrastructure projects; the rest was supported by investments from individual companies, as populations and markets grew. This meant that public aid and independent financing were enough to maintain a healthy economic cycle. However, in today's era of declining populations and diversifying lifestyles, which make for a market where demand is forecast to drop, if a single party invests in a digital platform to gauge demand-side trends, it is impossible to get a return. The only choice, then, is for the relevant parties to make efficient joint investments. In other words, unless we include a layer of mutual aid between independent financing and public aid, it is very likely that regional economies could grind to a halt.

In our explanation of digital platforms up until now, we have used the example of data integration platforms to obtain demand-side data.

Fig. 2-7. Ride-sharing service "Nokkal" and J-eID using services "Locopi" in Asahi, Toyama Pref.

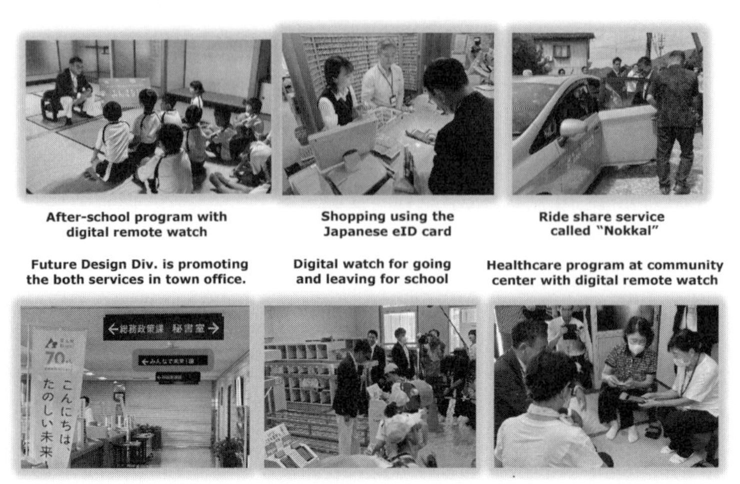

However, there are many other initiatives in the mutual-aid layer, including new public transport services that make use of automated vehicles and other digital technologies; cashless payment platforms for regional currencies; planning and implementation of regional sightseeing strategies using these services (destination-management organizations, or DMO); and the establishment of regional trading companies to open up sales channels for regional producers. There are many different social start-ups that could fit into this layer.

In the town of Asahi in Toyama Prefecture, for example, there is a project called Minmanabi, which roughly translates as "learning from each other." The project involves local senior citizens actively sharing skills like cooking and woodworking or natural experiences with local children. In return, the children help the senior citizens with digital literacy such as the use of smartphones. However, there was no transportation to take the children to the woodworking classes at the senior citizens' lumber workshops. And so, at the same time, the town launched the Nokkal service, a ride-sharing service that uses residents' private cars. This service also required help from a local taxi company. As the ride-sharing service got started, in addition to accessing these mutual educational initiatives locals also started using it to access health promotion services. Next, the idea was raised to somehow use the child support service for high school students. As a result, a virtuous cycle between the demand side and the supply side is beginning to take root. Surprisingly, there has been almost no impact on previous taxi demand.

This virtuous cycle between mutual aid and independent financing, as well as the role of public aid in supporting this cycle, could be called a new type of public–private partnership. We can now see that this is the goal to aim for. Let us again take a look at the three stages mentioned above—the warm-up stage, the rollout stage, and the expansion stage—to examine a hypothesis of where and how to begin.

2. Prepare (warm-up stage)

(1) The importance of diversity and inclusion

It has now been around eight years since the start of rural revitalization initiatives. Grants for promoting rural revitalization and other factors have contributed to the launch of many new initiatives in various regions. As exemplified by the corporate version of the Hometown Tax Program, we are also seeing an increase in support for these regional activities from corporations in more built-up areas. Issues remain, however.

While it is great that certain projects have reached the launch stage, it turns out that securing the necessary manpower prior to the projects' start is an issue that has not been dealt with effectively. For example, in many cases, after a project is finally set up, the majority of the people running that project turn out to actually be urban residents.

There are related issues after the projects have been implemented. Here, even though the long-awaited projects have been successful, the people involved have lacked the expertise to transform them into sustainable businesses. Moreover, the nature of the businesses is too different from that of traditional businesses for local business operators to take them over. As a result, the projects cannot be commercialized.

Due to these circumstances, quite often similar subsidized businesses are repeated in regions without taking root as established local businesses. Essentially, we are seeing the mass production of demonstration projects, with the process ending there.

We have identified three points that are key to changing this state of affairs. First, while it may seem to be a longer route, to tackle these issues it is important to improve regional diversity. Lifelong residents of these regions may not have the expertise to implement initiatives that are new and necessary. That said, rapidly amassing new people is not easily done, especially without connections. The aim is to gradually have a range of people with different skills and motivations visit the region on a regular basis. What will become apparent with experience is that, even if you receive a subsidy and start, if there are no people within the region with the potential to carry on the business, you will end up depending on people outside the region. When you start, if you secure local talent to be involved in the project, even if they can only engage in part of the project, or even if only one local person joins the project, it will make a huge difference to the project. At the same time, as large numbers of diverse individuals begin coming and going from the region, the number of so-called encumbrances ("unfair connections") will become negligible. As a result, it will be easier to avoid the stalemates that occur when those for and those against a project come up against one another when trying to start something new.

Later on, we will talk about how to attract these diverse individuals when you have not yet decided on a new project.

Second, it is important to fight back against the regional culture of subsidy distribution. People's perspective on money in regional areas is complex. In rural areas, people can live in relative prosperity without much money. Perhaps because they are aware of this fact, there is relatively little interest in increasing their own income. That said, they are interested in where money is going. In particular, more so than their own money, they maintain a careful watch over where and how money is distributed around them. Whereas in urban areas it is almost impossible to gauge how much money people are receiving and through what channels, in rural areas this is open for everyone to see. And so it is only natural that people in rural areas are curious about how money is being spent.

Thus, when a local government establishes a subsidy scheme to support local businesses, instead of an interest in what benefits the subsidized businesses will bring, people tend to look more at how much is being allocated to which sector and whether it will be fairly distributed within each sector. That said, no matter how large the subsidy is, simply distributing the money will not lead to new businesses or continuous employment opportunities. Instead, once the possible strategies have been narrowed down, the money should be invested in a focused manner over several years in specific businesses or business systems that have been given responsibility. However, rural areas that have focused on fair distribution of subsidies up to this point are not used to investing in a selected and focused manner.

Third, collaboration across different fields is essential for any rural revitalization effort. The project cannot only focus on food, or medicine, or health, or education, or transport, for example. Ultimately, it is important that multiple businesses are set up to handle these different fields, and that they can work together. If you set up a health business, can people access your service? If you promote healthy activity, is there healthy food on offer in the area? Taking this single example of health, positive outcomes will only arise when there are various initiatives involved. And so it is paramount that people with different backgrounds gather in the area.

A recent—but perhaps not lasting—phenomenon is outside people going to a growing number of regions to participate in a range of newly launched projects without planning to settle there. Although we are finally seeing diverse personnel come and go from these regions, there are almost no platforms or places where they can interact with one another. In fact, locals do not know much about who is coming and going, or indeed, about other projects. It is such a pity that there is not any space for these people to meet with each other.

Even if people from outside the region are given roles to play and places to belong through various projects, if those places are decentralized, people within the region have no opportunity to learn about the interesting staff members who have come to offer help. And as soon as the project ends, the outsiders return to the city.

In response, through teleworking grants and others, the Japanese government is promoting the construction of satellite offices in rural areas that everyone can share. This is something that we have previously introduced as the Inclusive Square in materials used in the Council for the Realization of the Vision for a Digital Garden City Nation.

It is essential to create interaction among diverse people from outside the region, and in turn promote mutual galvanization with the locals. When specific people are setting up specific projects separately, it is difficult for them to aim for the same summit and begin the phase that leads to the top.

(2) How to attract people

The warm-up stage, as we have briefly discussed above, is the stage in which to create an environment where various people come and go from a region in preparation for change. Preparing the region through this activity is essential to promoting new investment cycles and creating highly productive new businesses. If you try to suddenly set up a single "hot" business without warming up the region beforehand, it will be difficult for the business to take root. This is why you must first ensure diverse people come and go from the region to spark lively exchange. This process is essential if you are aiming to transform regional economies and lifestyles. We have identified three important strategies to facilitate this process.

(i) Gently increase involvement

First, rather than suddenly asking people to move to the area or get deeply involved, it is important to increase their involvement gently.

For example, one local government is slowly and deliberately working to increase points of contact with outside people. First, they encourage people to join the Hometown Tax Program, and then host a thank-you event for frequent donators. Next, through a contact point, they encourage people who frequently attend these thank-you events to visit the region before finally encouraging them to participate in local businesses. This kind of strategy will get more people coming and going before you know it. The town of Kamishihoro in Hokkaido is one example; here, as with all regions that have succeeded in increasing interaction between people inside and outside the region, the key has been to gently and gradually increase the variety and number of regional points of contact. The people from outside must be more than simply tourists, but not quite citizens. Whether they are drawing fans of the region or others, the region in question must try to gain as many people that fit into this mold as possible. We could call these people "connected minds," or even the "tourist population" or the "exchange population." Almost without exception, local governments that are intentionally working to create both internal and external supporters have succeeded in attracting a diverse range of human resources.

The point is to keep their involvement with the region continuous. For example, one might become interested in Tokushima Prefecture's traditional Awa Odori dance. At some point, rather than watching the dance during the festival, that person might begin to enjoy taking part in Awa Odori preparations. In turn, they could then take part in other projects in Tokushima Prefecture. Gently increasing the number and variety of contact points with the region is paramount. This could be through the Hometown Tax Program, famous sightseeing spots, local festivals, or local cuisine. It might mean attracting people interested in the history of the region, or those who frequently take part in local nature experiences.

(ii) Create roles to play and places to belong

Once you have succeeded in gently increasing regional involvement, the next step is to intentionally create easy-to-understand roles to play and places to belong.

Perhaps surprisingly, job titles will play a significant role. Although not all people from outside the region care about their job titles, the locals should proactively verbalize and define their roles to play by giving them such titles as project advisor, project manager, festival prep team, nature experience instructor, or so on. That will give them a good reason for coming back to the region. Moreover, to ensure that these people do not simply go back and forth between their hotel and the worksite before heading back to the city, you must create places where they can belong, whether it is a permanent office or a cafe. It is crucial that there be no delay in creating roles to play and places to belong for the outside people who have begun to establish themselves in the region.

The various subsidy programs offered by the national and local governments can help create roles to play and places to belong for people both inside and outside the region. This is what ultimately will help transform the region. But when the number of eligible projects increases, it is impossible to subsidize all of them through selection and concentration. Conversely, if finances allow, an interesting option is to launch various initiatives as a way to actively increase the number of outside people involved with the region, regardless of the subsidy.

(iii) Engage in purposeful communication

Without exception, regions that have successfully launched projects and succeeded in attracting outside people to the area have engaged in purposeful communication. Conversely, the reason many regions struggle in the initial stage is a complete lack of communication. When people start something new within the region, if people outside the region are not aware of it, then no one will take an interest and come.

This communication requires a careful strategy. Its purpose is not for the region to become well-known nationwide; rather, the aim is for people interested in the region to introduce its charms to others around them. Whether it is a tourism portal website or a website to introduce the region's initiatives, the important thing is to make sure the content is aligned with this aim. Simply showcasing the region's famous sightseeing spots or the results of regional projects is not enough. Websites that are updated once every six months will not get repeat visitors.

A regularly updated website will offer new discoveries every time someone views it. Or what if the website shows surprising and compelling photos from the region, which people will want to share on social media? In this way, the region must engage in purposeful communication that makes the receiver want to share the content with people around them. It is from this stage that smooth involvement with the region can begin. Even though this is just the warm-up stage, it is important not to neglect communication of the region's charms.

This communication also brings valuable secondary effects. That is, it makes people within the region aware of the fact that people from the outside are taking an interest. Whether it is undertaking a new project or reforming lifestyle services, seeking spontaneous change from within a region is not easy. However, if regional initiatives are picked up by nationwide news outlets, or people from outside the region take an interest, things begin to change. Regional projects that were once someone else's business become more of a personal concern.

Let us now sum up the above initiatives.

a) Gently reinforce points of contact with the region
b) Create numerous roles to play and places to belong for people from outside the region
c) Reinforce purposeful external communication

Intentionally working on these three initiatives will gradually create an environment in preparation for regional change.

Rural revitalization could be compared to summiting a mountain as follows:

(i) Create numerous climbing routes to the summit and take care to avoid steep cliffs and trails

(ii) Assign titles to the mountaineering team and create mountain huts where they can take a meaningful break
(iii) Send frequent updates on the mountain-climbing expedition and the mountain itself to external parties

3. Identify a key business (rollout stage)

Once the warm-up stage is underway, the rollout stage is next. While there is no one right answer, we recommend tackling the rollout stage with what we call an "inverse T-shaped model." With this model, you first identify a key business and then prepare a digital platform to support it. Why, then, is it important to identify the key business? And, at the same time, why is the creation of a digital platform so effective? We will go through the necessary points in the relevant order.

(1) Organize the governance framework

Let us begin by taking a look at what a general partner, or GP, is. Unlike a limited partner (LP), which is an investor with limited liability based on the size of their investment, a GP has unlimited liability. In simple terms, when setting up a business, a GP is at the center of that business. To pose an extreme case, if the business went bankrupt with debt, the GP would be responsible for repaying that amount to the very last yen.

If you want to set up a sustainable business, someone has to take full responsibility for that business. Without this responsibility, the business will not be able to continuously raise funds, among other things. Many subsidized projects begin with a false sense of security, going forward without a proper GP in place, because they rely on the idea that the subsidy will save them. This is one of the most common reasons why many projects fail to develop into full-fledged businesses. A clear liability framework, or governance framework, is therefore essential. This framework must be created when deciding on the key projects to launch. When setting up subsidized businesses, those involved will often draw up an implementation framework. But these diagrams do not clarify who is ultimately responsible for the business. Clarifying a liability framework to back up this implementation framework is key.

Recently there has been an increase in the number of large companies and supporters who are willing to invest in rural revitalization projects as LPs. Donations through the corporate version of the Hometown Tax Program are also beginning to increase rapidly. This is no doubt because investing in businesses that could have a positive impact on society can make large companies more attractive to investors

Fig. 2-8. What is a general partner?

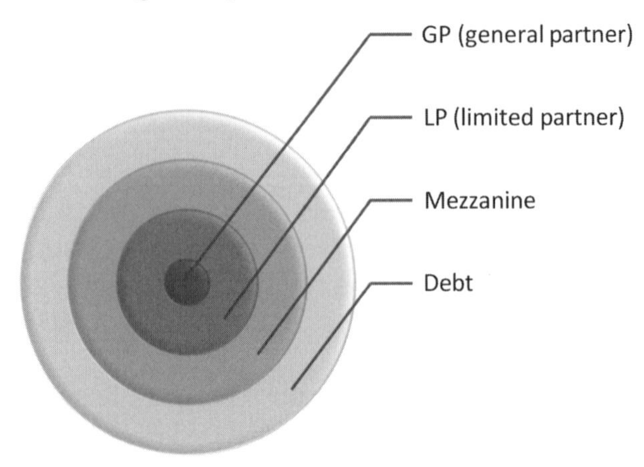

Fig. 2-9. The trend in corporate version of hometown tax donations

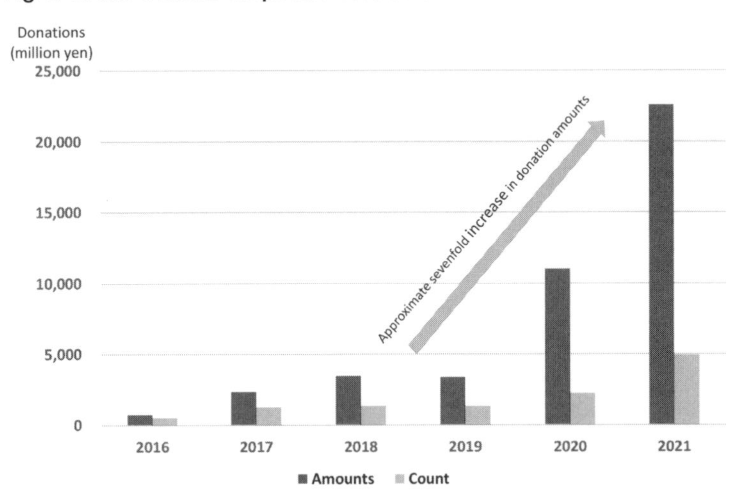

and in turn support their investor-relations activities and maintain their stock value.

While this in itself is very positive, the problem is that regions are launching businesses without a GP, basing a false sense of security on the fact that they have successfully raised funds through LPs. But without a GP, if the subsidy runs out, for example, it will be very difficult to raise the funds for the next stage of the business. As a result, it will end up being a one-time project. Of course, this is not what the large companies who have invested in the business want to happen. But it is also difficult to get these large companies on board as GPs. The challenge, therefore, is securing the GP.

A related issue is that there are very few people or organizations who can take on the role of GP within the region itself. For example, if someone who has developed and inherited a strong local family business is asked to take on a new initiative as a GP, there is too great a risk that their valued business could suffer. Furthermore, very few young people in rural areas will be willing to engage in an entirely new business.

That said, it is important not to think that rural revitalization will be a challenge simply because of the lack of young management personnel. Even in urban areas, there are no prestigious fund owners or business managers who were reputable GPs or managers from the outset. In this sense, the lack of young management personnel is not unique to rural areas. Management qualities are not innate, but built up through experience. And so, rather than worrying about the availability of personnel, it is more important to look at whether there are environments in place to enable personnel to take on new challenges or that can encourage them to do so. However, the problem with regional businesses compared to, for example, tech start-ups in big cities, there is a strong likelihood that GPs will take a loss.

Even though investing in tech start-ups that handle the most advanced technologies carries significant risk, the returns can be massive. This means the disparity in possible returns between GPs and LPs is substantial. The problem is that social start-ups are not very profitable. With such small profits, even if you tried to increase a GP's share, the difference would be minimal, and so in the end, despite the GP taking on more risk, GPs and LPs would see very similar returns. Clearly, the risks are not balanced with the potential benefits. And so no one puts their hand up to become a GP, and no GPs can be developed. To tackle this issue, instead of looking for people that do not exist, it is essential to create environments that are likely to cultivate GPs, and intentionally go about creating frameworks which are advantageous for them. The creation of such a framework for the key business is essential.

Fig. 2-10. Inverse T-shaped model

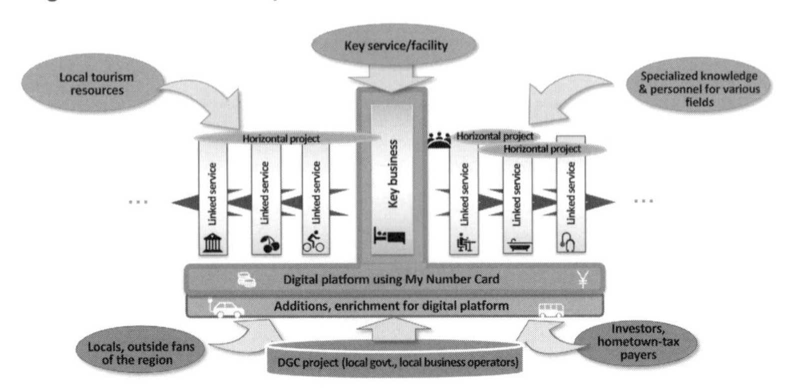

(2) Identify a key business

To begin with, the aim must be to intentionally choose and set up a single key business in each area. At the same time, you must create a digital platform at the center of this business to anchor it. This is what is encouraged by the Council for the Realization of the Vision for a Digital Garden City Nation with the prime minister present.

Let us say that there is a single commercial retail and accommodation complex within a region with an excellent ability to attract customers. While for locals it is a separate world which they rarely visit, this complex is frequently overbooked. In this case, it would be great to have a system in place to efficiently transfer the overbooked customers to accommodations in the surrounding area. It would be even better to be able to introduce a digital platform that could host a system of regional loyalty points and rewards and a reservation management system, and to be able to use loyalty points and discounts as an incentive to attract overnight guests. Mechanisms like these must be set up for the key business without compromise. Once this system starts functioning, whether for the sale of local specialties or the provision of activities and experiences, the same platform of loyalty points and discounts could be used effectively to extend the time people spend in the region.

It is, however, important to avoid launching multiple key businesses simultaneously. The reason is as follows: the locals may have given way on the selection and concentration of investment in the key business for the sake of the region's future, but if you launch three or four business at the same time, you would just be following a conventional distribution pattern.

Although there is no need to communicate topics like this to the relevant parties on every occasion, those setting up a business must work to move away from the conventional "distribution/consumption" culture to the "focused investment" culture (selection and concentration) with a constant eye to generating an investment cycle within the region. Those involved must take care not to spark a debate about the distribution of subsidies.

So what is important when choosing and launching a key business? Let us take a look at three points.

(i) A business that is mindful of the region's main issues and whose significance is easy to explain

First, whether it is disaster prevention in specific areas or support for specific types of underprivileged children, no matter how important the issue is, if it is too specific a project, it can be difficult to gain the support of the whole region from the outset.

As such, the best way forward is to select an important business that can gain the support of a comparatively large proportion of the region and grow with ease, such as bolstering sightseeing and travel, improving public transportation, or improving healthcare or childcare.

(ii) A business with a complete framework in place

Second, a key business cannot move forward without a GP. Each business must have a thorough governance framework in place and be able to independently ensure the flow of funds. It does not matter if this framework gradually takes shape as the business unfolds. While this may sound difficult, it is important to create a business framework that allows for the use of both subsidies and external funding to grow the business using your own initiative.

With many public projects, although the details may appear to be reasonable, a look under the surface reveals a poor supporting structure. In many subsidized businesses, although the nature of the business is determined through various repeated discussions, when it comes to deciding on who is responsible for implementing the business operations, the only choice is to seek such people through open recruitment.

Fig. 2-11. Disclosure, governance, and financing

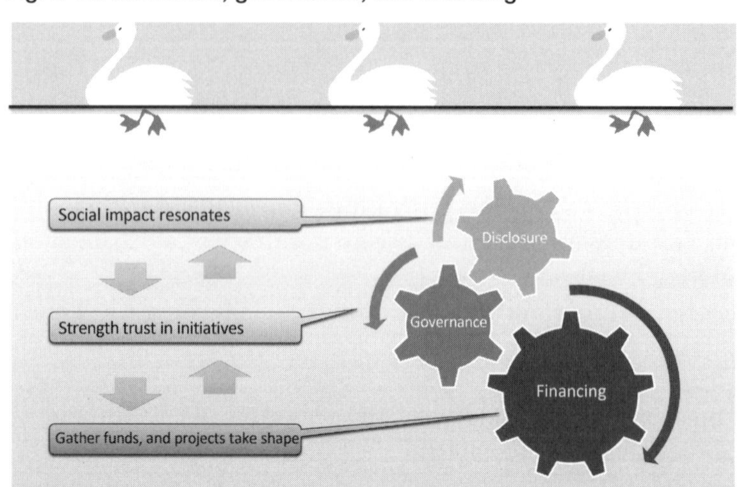

The important thing, therefore, is to ensure a strong structure beneath the surface.

This is one of the major differences from launching a business in the private sector. In the private sector, while of course the nature of the business is open to discussion, greater priority is placed on who is responsible and where the money will come from. For some reason, public projects place more importance on the nature of the project than on who is responsible. While on the surface there may appear to be a group of strong and elegant swans, a look under the water shows just how weakly they are paddling.

To avoid this situation, when setting up a business you must ensure that each entity has three fundamental elements in place: governance, disclosure, and financing. As above, any entity without a liability framework will not be able to generate funds. To raise these funds, it is essential that you proactively communicate the nature of the business to potential investors. Financing decisions will depend on these communications.

Although it may sound complex, we can redefine financing as the creation of relationships with people through money. This could be individuals who want to contribute to the business, those who want to invest, or those who want to play a central role in business management.

There will no doubt be people with various ideas and hopes in and around your business. And so, when launching a business, you must build suitable relationships with those people and carefully select the methods you will use to raise funds.

Really, creating a business framework like this is the job of an experienced management professional. In some cases, it may be difficult to access all of these elements in a single region. This is why a network of business accelerators is key for any region. Business seeds, GPs, and business professionals are all closely related.

(iii) Creating a framework that enables the public and private sectors to support the GP

Third, there must be an adequate public–private partnership in place. This does not mean that a subsidy should be granted. Even if the local government does not provide funding support, if it has confidence in your business, this can have a significant impact on the success of the business. There are many things a local government can do to help the key business, be it clarifying the relevant regulations, helping with labor procedures, or lending a hand in building relationships with locals. Elsewhere, it is important that you maintain a sense of distance from the GP so as not to unnecessarily interfere with snap management decisions based on market trends. Rather, this might be the job of LPs or locals who are looking forward to the business's success.

When we interviewed the business operators who had gathered in Kamishihoro and asked them why they had selected that area, many said they chose the area because it was "fair." Almost every morning, the town mayor visits the office of each business to check whether they are having any issues and whether they are using their subsidies efficiently. On the other hand, while the town works hard to arrange business opportunities, it does not try to attract businesses by offering financial support. Instead, the town treats every business that has taken an interest equally. This is one of its key characteristics.

Moving forward, it will become increasingly important to use mutual aid and work together to develop businesses required for town development. While the shape of public–private partnerships will differ depending on the nature of the business, there is no doubt that a business will not succeed with only public or private efforts. The key is to ensure that each party supports the other like a pacesetter, without depending on them too much or staying too far apart.

(3) Digital platforms

The second element that makes up the inverse T-shaped model, and that must be set up at the same time, is a digital platform. Here we take a closer look at what a digital platform should look like. Let us first take a look at some examples of digital platforms that support key businesses.

(i) Set up an authentication and payment platform using My Number Cards in each region. The key business here is a cross-reservation system among accommodation facilities that features loyalty points and discounts. The goal is to extend this cross-business collaboration to the sale of local specialties or the provision of unique experiences, and in turn have local businesses work together to expand the regional tourism market.

(ii) Create a digital platform that can manage automated and other on-demand public transport alongside demand data. The key business here is a new education-related business such as mutual education. The goal is to extend its application to other childcare and nursing-care support programs, and in turn create a virtuous cycle between public transport and demand.

(iii) Distribute devices with easy-to-use pre-installed apps to senior citizens. The key business here is a lifestyle support service for senior citizens. The goal is to gradually extend the business to the digital transformation of administrative tasks in terms of cooperation from local volunteers (residents' association, etc.) for disaster preparedness and other purposes.

Whether it is the authentication platform using My Number Cards, the platform to gauge demand for public transport, or the shared digital platform for senior citizens, the key is to ensure it is equally available to like-minded businesses within the region at a low price.

Even if, for example, you create a digital platform for a regional currency, it will not be much use if a competitor prevents you from bringing together data on spending history in the region. Moreover, if a system to gauge transport demand is only available to specific transport operators, it cannot be called a regional platform. It is also best to avoid shared devices that restrict users to specific telecommunications providers. In any case, the most important thing is to make the platforms available to people trying new lifestyle services at a low cost. When the key business succeeds, these digital platforms are essential to quickly bringing the second and third businesses to fruition.

To date, some rural revitalization initiatives have undoubtedly developed into positive businesses. However, there is no benefit to the area as a whole when that business is the only one increasing its productivity and profitability. A key business is one whose own success leads to the success of ensuing businesses.

Let us say, for example, that you have created a shared platform that enables the My Number Card to be used as a means of identification, and where scanning the card in a store or at a counter gives the owner access to digital gift certificates and other loyalty rewards. While this could begin with loyalty points and discounts across accommodation facilities, it would be even better if its application could be gradually extended to local specialties, events, and services before moving on to things like weekend activities for locals, and ultimately healthcare tourism. Such a mechanism has the potential to create a virtuous cycle that both increases customer activity and enhances customer services.

If you create a digital platform to manage data on public transport and demand for transportation, and successfully establish a supply and demand cycle using the key service business and the digital platform, this platform can be used to support the next business, as happened in Asahi. If your digital platform is based on devices for senior citizens, after successfully popularizing the initial service, you can use the same device for the next service and keep distributing more of them. This is how you can use a key business to develop businesses that follow.

(4) Target area (the area theory)

Although there is already a great deal to digest, allow us to add another important point of discussion: the area theory.

Upon setting up a digital platform, collaboration between local governments over a broad area is an important step. This is because the key business and the businesses that it generates must not be limited to a specific region. The new service businesses that branch out from the initial key business can only do so based on the scope of usability of the digital platform. Therefore, the target area for the digital platform must be a key consideration at the launch stage. This area theory comprises three important points.

First, you must prepare for a broad reach. If the scope of the platform is limited to a specific region, the growth of sightseeing services that can be built on top of the platform will be limited from the start. It is important to consciously establish an economic zone where you want to achieve revitalization and independence. For example, in the case of Mie Prefecture, the economic zone should cover not just Taki, but all five beautiful towns; and in the case of the Izu Peninsula, the economic zone should cover not just the city of Mishima, but the federated municipalities of the peninsula. To ensure that services in the economic zone have significant market potential, the platform must be set up with an eye on collaboration over a broad area.

The area in question must be large enough and have sufficient density of demand that investors can expect significant returns on the various lifestyle services they choose to invest in. In terms of population, the ideal number would be upward of 100,000, but less than 300,000. A look at the independent regional economic zones being discussed as part of the government's National Spatial Strategies will provide more details. However, in some cases it will be difficult to discuss the details of broad-based collaboration from the outset. In this event, one option would be to begin on a small scale and gradually expand.

To the second point, the adoption of a uniform area brand is also paramount. In a single area with disparate brands—like Niigata Prefecture with the Echigo-Tsumari Art Trienniale, the Snow Country tourist zone, Minamiuonuma, southern Niigata, and Uonuma-no-sato —the multiple brands might hamper external recognition of the area, despite there being a great opportunity to increase recognition. (To be clear, in Niigata, through the accumulation of many years of steady activity, the Echigo-Tsumari Art Trienniale and the Snow Country tourist zone have succeeded in garnering a high level of external recognition.)

Of course, the background behind the development of every facility and project is different. It is unusual for everything within an area to be seamlessly linked through a keyword like the Art Trienniale is in Echigo-Tsumari. In some cases, there can be a lack of uniformity within

Fig. 2-12. Link between individual businesses and the area (Area theory)

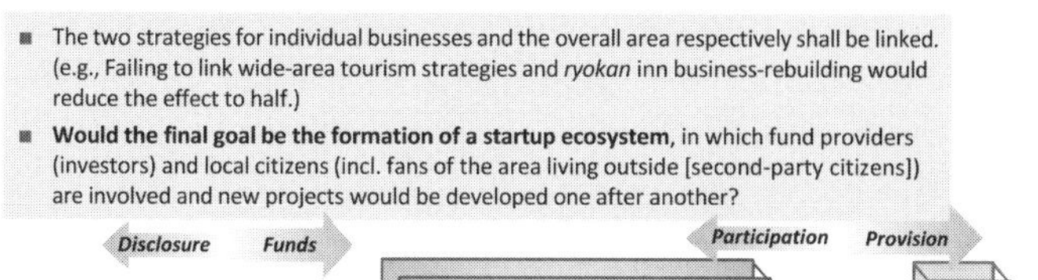

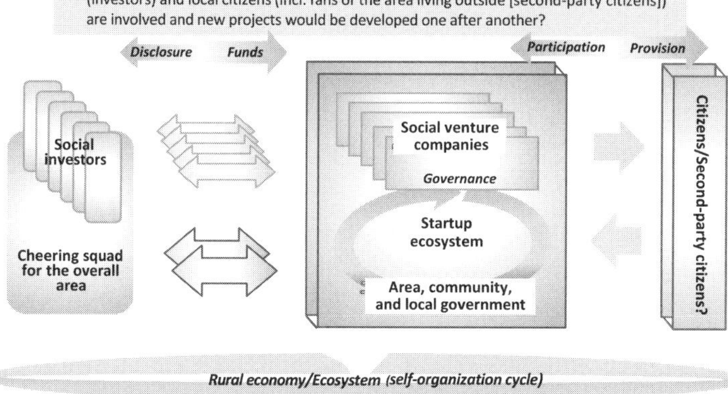

a single region. For example, there might be two adjacent facilities with contradictory branding, where one emphasizes a "tourist zone," and another highlights that it is a "town for locals." Places tend to value their own individual reasoning and purpose. Although this approach may help to maintain sense of equity in each region, it is not beneficial for selected and focused investments. Delineating an area for broad collaboration and thinking comprehensively about the brand to create can be a more challenging task than you might think.

Third, once you begin creating an area brand, you must incorporate the businesses of existing operators in the area. After settling on an area, if you only focus on certain new demonstration businesses and do not connect with local lifestyles, your efforts will not lead to improved productivity for the local businesses, and will not boost local employment or improve its quality.

For example, it is often the case that the people devising broad-based tourism strategies do not interact at all with the people responsible for helping the struggling managers of accommodation facilities. The former group puts together its ideal regional strategies without knowing the struggles of local accommodation providers, while the latter gives pointers to struggling managers on how to obtain subsidies without knowing the nature of these broad-based tourism strategies.

Even outside of tourism, if the people examining area strategies do not take into account the management indicators of each business operator and incorporate each business, regardless of its commercial potential, the key business will end up as a mere demonstration of a broad-based strategy.

In this way, it is essential that you take the management issues of existing businesses on board as you work to grow your key business into other businesses. Each region must ensure that the strategies of individual businesses are in line with the overall area's strategy. As we said at the start, this is the process whereby you climb the summit together. As one path up this summit, you could use the formulation of Comprehensive Strategies for the Vision for a Digital Garden City Nation for each area, which is something that each local government will do moving forward.

With a key business at the center, if individual business strategies can easily be linked to the business based on a digital platform, this will naturally lead to the creation of a regional start-up ecosystem. Strategic thinking toward this goal is what regions need. The spotlight now is on the strength of communities and regions, and this strength is exactly what is required to manifest strategic thinking.

4. Considering what is required for expansion (expansion stage)

Once the key business is on track, the next thing to do is successively launch new businesses. This is called the expansion stage.

It is during this stage that the diversity and the roles to play and places to belong that were cultivated in the warm-up stage begin to come alive alongside the digital platforms that will facilitate synergy. Speed is essential. In this stage, the aim is to transform a region that was previously resigned to or satisfied with staying the same into one that generates successive social start-ups that tackle local challenges. Going too slowly, however, could give the region space to fall back into the same old rut.

To avoid this, there are three things you must be conscious of.

(i) First, there must be a business accelerator who assists the GP and supports the creation of various businesses. If you are only focusing on the key business, you can deal with it on an individual basis. You may even be able to gradually build up the governance framework for the key business as it begins. However, when it comes to organizing the second and third businesses, for example, it will not work unless you increase the number of venues for interaction with management professionals who will assist in establishing the business environment for each one, interfacing area strategies and individual company strategies, and raising funds. There does not necessarily need to be a business accelerator in every region. One option could be to create a nationwide network of accelerators and have them commit to working with a region as and when required.

(ii) Second, the environment must be conducive to developing potential GPs. At the Trailblazers Conference, Toshiki Abe from Ridilover called this a "kelp forest"; that is, a safe environment for individuals to grow. While the best way forward is to gradually work to create this "kelp forest" from the warm-up stage, a few other aspects of the process should be noted. The environment you create must allow potential GPs to find a place to belong within the region while being neither too conspicuous nor too hidden. This "kelp forest" should also have a shared target; that is, a summit to aim for. It does not matter whether your initial business is related to tourism, healthcare, or education. Ideally you will create a "kelp forest" where GPs who have ideas for town development will quietly but naturally gather.

(iii) Third, town development must utilize well-being indicators and draw upon data to measure the social impact of the project throughout the area. When the project begins in earnest, objective data must be available for different discussions and the adjustment of interests. It will also be important to link and share the data obtained via your digital platform to discuss the social impact of your initiatives and local well-being with locals, outside supporters of the region, and businesses.

Below are simple descriptions of each key point.

(1) The necessity of a business accelerator network

As explained in the discussion of GPs above, when setting up a business, you must return to three fundamentals: governance, disclosure, and financing. Without a liability framework in place, it will not be possible to generate funds continuously and sustainably. It is axiomatic that a business must be able to continuously generate funds if it is to continue. Subsidies alone will not be enough. To raise these funds, it is essential that you proactively communicate the nature of the business to potential investors, and at the same time set up a liability framework to gain the confidence of these investors. Failure to do so will make the business reliant on subsidies that it will have to apply for repeatedly.

But a wide range of expertise and experience is necessary for such fundraising. You will need help from professionals in various fields with expertise in legal affairs, labor affairs, accounting, fundraising, human resources, office environments, presentations, and much more. It is unlikely that you will find people with all these skills within the region in question. Moreover, it is almost impossible to bring together people with all this experience in each region.

In the US, which has a flourishing start-up scene, venture capital and angel investors help to support start-ups, but there is also always a business accelerator behind the scenes. Business accelerator programs are made up of experts in speeding up growth processes; they develop fresh start-ups into full-scale businesses that can aim for an initial public offering.

In some cases, to provide start-ups with this expert support, business accelerators can provide shared offices and various other functions from the start-up's launch stage. Because we do not have a hub for start-ups in Japan like Silicon Valley in the US, we will need to create a nationwide network of business accelerators who can attend and support regions that have reached the relevant stage of development. This is one task that Japan will need to focus on moving forward.

(2) Creating an environment conducive to the development of GP candidates (kelp forest)

Unfortunately, people do not suddenly decide to accept the role of GP on a given day. A gradual process must take place in each region. However, when what needs to be done is not yet clear, it can be difficult for potential candidates to immerse themselves in the region straight away.

There must be opportunities for these people to gradually take part and help with regional businesses. Perhaps this can then lead to frequent support for a specific project. They might then take the lead in a special limited-time project. By gradually helping with regional services, such as by teaching educational courses, ultimately they will end up frequently coming and going from the region. While there are many ways to go about facilitating these interactions, you must create opportunities for people who might one day decide to help launch a business to quietly and naturally blend into the region. Toshiki Abe from Ridilover has called this kind of space a "kelp forest."

Although not all of these people will become GPs, it is important to create an environment where they can find places to belong within the region without being too conspicuous or too hidden. If outside people stand out too much, it might become difficult for them to stay there. Without a forum to continuously engage with the region, however, they cannot get to know the area or observe any new trends. The "kelp forest" allows these potential GPs to stay involved with the region with an appropriate sense of distance. If possible, it can be very effective for these people to build some kind of rapport among peers in the same situation in advance.

One way model for creating this kelp forest can be seen in the structure of Setouchi Works in Mitoyo or the Multiple Work Cooperative in Ama. Rather than having people who have decided to take part in rural revitalization immerse themselves in a specific regional business straight away, it could be beneficial to intentionally create something like an informal gathering place—one that is not so obvious—where they can try out different things and gradually strengthen their interactions with the area. Creating such a place will ensure that when there is a chain reaction of events following the success of a key business, a new investment cycle will quickly grow.

(3) Use of well-being indicators and collection of data to measure social impact

The final point is the importance of well-being indicators and other data. This is important for three reasons:

(i) The resolution of regional social issues requires initiatives across multiple fields.
(ii) Data pertaining to local lifestyles will play an important role in the adjustment of interests and the incorporation of various people. This is where well-being indicators will play a key role.
(iii) Organizing these different elements and measuring the social impact of various businesses on the region can ultimately link the region's successful businesses to global-impact investments.

Let us now break these down in order.

(i) Town development requires initiatives across multiple fields

Let us use the health-promotion field as an example. One common idea with recent projects has been to use wearable devices to measure exercise-related data in a fitness class for senior citizens. Naturally, business KPIs include targets relating to the health of senior citizens and the extent to which medical costs are reduced. If we take a more detailed look, however, how are senior citizens expected to attend fitness classes in a region with limited public transport? If they stay at home, how can they receive remote guidance? Even if these fitness classes succeed, how can you maintain the health of local senior citizens who have poor diets? Even if senior citizens are physically healthy, can they stay healthy if they are lonely and struggle psychologically?

With rural revitalization, if you are trying to create a region where people can live happily and with vitality, whether by addressing health, education, public transportation, or disaster preparedness, tackling a single field alone will not allow you to achieve your ultimate target. Without public transportation, businesses relating to health, education, or disaster preparedness will not succeed. Elsewhere, disaster preparedness businesses can only succeed with initial training, while public transport businesses will not work without demand. Town development needs multiple fields working together and contributing to one another for everyone to reach the summit of the mountain.

As such, mutual collaboration between initiatives in different fields is paramount. To date, many towns have indeed succeeded with initiatives exclusive to certain fields, such as public transportation, education, or health. But this was when populations were rising. This meant that as the supply side increased its volume and options, it could pay a certain amount of attention to initiatives in other fields and naturally create synergy.

However, as populations have begun to fall, as we mentioned at the start, it is not possible to increase volume or options. In this context, it is essential that each field meet the needs of the others, be it the transportation required for medical care or education, or the demand required for the revitalization of public transport. Each field must accurately gauge those needs, and rather than waiting for natural market adjustments, must consciously and proactively create those connections based on the available data.

The challenge here is to facilitate communication between those in charge of each service business and to ensure that everyone shares the same town-development vision. In many cases, local businesses in the same industry share the same circumstancesfate, yet are fierce rivals, so when it comes to business, they will not agree on collaboration straight away. Like the sometimes-difficult relationships between heads of neighboring local governments, relationships between people in the same industry or among local corporate managers are rarely smooth sailing.

This is where the capability of the region is questioned in terms of naturally generating collaboration between multiple fields; in other words, initiatives on the "mutual aid" layer. The local government alone cannot be responsible for coordinating these relationships and balancing different interests in different fields, since bureaucrats work in their own silo.

(ii) Data is essential for adjusting interests

How, then, do we go about promoting collaboration toward a shared vision and business goal despite the potential lack of smooth relationships? When looking at past examples, the answer appears to be in the use of data.

Let us use the example of a local shopping street trying to start a new project, something which typically is never straightforward. The suggestion for the project was to set up cameras along the street to obtain data on the flow of people. However, both advocates of the business and those who were more cautious about the business had their own valid opinions. As a result, it was difficult to enter into discussions on the finer details, and as a result there was no choice but to abandon the business. But what if there was already data that showed that the flow of people along the shopping street was definitively decreasing, and that their ages and attributes were changing? Of course, this data will not instantly lead to the launch of a business, but at the very least it will allow full-scale discussions to take place, unlike the above hypothetical. Communication can be difficult, and may be influenced by preconceived notions. For example, you are more likely to disregard the opinions of people you dislike and agree with the opinions of those you do like. Unless discussions begin from a neutral standpoint based on available data, organizing discussions to win over locals can be a challenge in itself.

Another important variable is the involvement of citizens. The local business community and economic organizations have always worked and will work for the benefit of the region. However, when it comes to their own businesses, moving discussions forward can be difficult. They are well versed in who does and does not work well together, but they cover up these difficult internal relationships when discussing what system to use to launch the new business. Once that system is selected under the "selection and concentration" principle and people who do not get along are to be involved in establishing it, it is difficult for the organizers to obtain an agreement directly from those people.

Who, then, can generate collaboration among local business operators? We think that citizens—their voices—hold the key. The term "citizens" here does not necessarily mean those who are on the local resident register. There are some people who are not on the register but are making huge contributions to the region's future; conversely, there are some locals on the register who have little interest. The key is to value the people who love the region, even if they showed up midway through the process.

Again, data is essential for ensuring that people join businesses that will help shape the future of the region. This is because no one will listen if a citizen or supporter outside the region just talks about their own ideals and beliefs when discussing the region's future. Show them the data, and show them the plans. When local business operators ask questions about the data and plans, discussions can begin. Without data, it will be difficult to generate discussions among local business operators, or between citizens and business operators who have previously had little interaction. Conversely, data can lead to discussions covering a wide range of angles. Using data to talk about the region is essential when considering its future and the potential for expanding the key business across the region.

The concept of well-being and well-being indicators will also play an important role.

The Digital Agency and the Smart City Institute Japan are currently working together to create a system for the measurement of well-being indicators. This system is made up of objective indicators from actual statistical data and subjective indicators from surveys. Together, the two organizations have created a system that allows all municipalities in Japan to download the required statistical data for free. For the surveys, the two organizations have made survey sheets available for all municipalities to use. Other work is currently underway to create simplified

Fig. 2-13. Components of well-being indicators

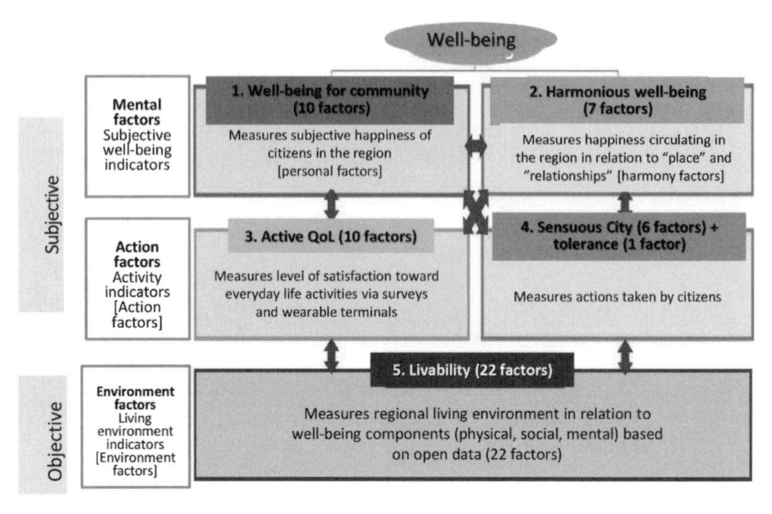

Source: The Smart City Institute Japan
Council for the Realization of the Vision for a DGC (7th meeting), Minister Makishima materials

versions with fewer questions and to develop online surveys.

What citizens expect from town development is not the improvement of business profits and productivity. In order for citizens, businesses, and local governments to have common grounds for discussion, talks must be framed around how to enhance the livability of the town and the happiness that living locally provides. While we will leave the details for another occasion, discussions between business operators and citizens using these well-being indicators could play an incredibly important role in boosting the strengths of regional communities and building momentum toward a common objective.

(iii) Linking a business with global impact investments

To finish, we would like to touch upon something of a slightly different nature.

Business continuity requires continuous fundraising. Businesses that are working to improve productivity cannot move to the next step by looking at annual sales alone. However, if there is not enough capital within the region, the business must bring in external funding after setting up a thorough governance framework. This will be a long-term challenge moving forward, but for rural revitalization to succeed, it will access to continuous financing.

Of course, regional services will not be able to obtain the same funding as tech start-ups, unicorns, and other projects with high risk but high returns. This is because the finance industry prioritizes relative returns over definitive returns. For example, if a project with an internal rate of return (IRR) of 10 percent is presented alongside one with an IRR of 6 percent, the latter will not even get a look. However, financing will not be an issue if you promote a project with an IRR of 6 percent alongside others that only offer 5 percent. Still, you are unlikely to find regional service businesses that can rival the returns of a regular investment project with a high IRR.

Impact investing works slightly differently. Usually, people invest with a view to getting a return. Impact investing puts more weight on investing in projects that can have a positive impact on society. In the global finance industry, there is an overabundance of capital. If investors only seek businesses with high returns, it is hard to find a sufficient number of investees. However, this excess capital cannot simply be left to sit, and this is where impact investing comes in. Even if returns are low, knowing that the investment is having a positive social impact can be enough. This excess capital supports environmental, social, and governance (ESG) investing and sustainable development goals (SDGs).

Moreover, companies are finding that their approach to ESG investing and SDGs, as well as their level of social contribution through impact investing, is beginning to have a significant effect on their stock prices. This is why it is important to link the demand for funding in regions to the impact investing market.

Two conditions are necessary to make this connection. First, you must describe the social impact of each business on the region using quantitative data. Second, you must already have a certain level of financing.

To get the investments you need, you must ensure that potential investors can access the information required to make their decision even if they cannot visit in person. We would like fund managers, corporate venture capital managers, and managers of companies seeking investees to visit the region and make their own assessments of the business. However, many potential investors, such as fund partners or executives from the company considering investing, will not find time to visit the region until they understand the nature of the project. It will be almost impossible to access the necessary funding without data-based descriptions that can persuade potential investors.

As mentioned above, if you truly want to reduce medical fees and boost the health of senior citizens, for example, you must approach your business from a comprehensive range of angles, combining fitness classes, public transport, and even diet. In many cases, you might need to organize mutual relationships through a logic tree or other mechanism, and conduct demonstrations once you have acquired a certain amount of data.

Second, you must also determine the scale of financing needed. For potential investors, the time and effort required to review and examine a 5-million-yen investment and a 500-million-yen investment is the same. For example, if you estimate the personnel expenses required for careful examination of the investment and the post-investment monitoring process at 5 percent, it is only when the scale of funding reaches one billion yen that you can allocate 50 million yen to personnel. As a result, the interest of managers of impact-finance institutions will naturally shift to other investees where they can invest a larger amount of money with the same time and effort.

If you are running a business that is highly beneficial to society, but you lack the data to explain your activities, or are not used to explaining your social impact, discerning investors will need to allocate excellent personnel to the business. But if you do so with a 5-million-yen project, at 5 percent for personnel expenses they will only be able to allocate 250,000 yen, which at best will only cover actual costs. To be specific, it is only when you can raise more than one billion yen that you will be able to have the investor use outstanding personnel for the project.

Of course, regional projects that can instantly raise one billion yen are few and far between. However, whether it is a health-related business or a similar business focusing on broad-based collaboration, it can become easier for investors to commit if you can incorporate the businesses into your social-impact narrative. Opportunities like these may arise in the expansion stage.

There are currently not many regions with projects that have gotten to the point where they can strategically raise this level of funding. One of the goals of the Digital Garden City Initiative, however, is to create an environment for the continuous development of social start-ups based on the commitment of continuous-impact investments. Individuals who have succeeded in launching key businesses and digital platforms and are at the center of the area's investment activities will play a key role moving forward. Together with local governments, they must bring together businesses that have a positive social impact and strive to link them to global impact investments.

Summary

To conclude the chapter, let us summarize what we have discussed so far.

The ideal way forward for rural revitalization is to focus on generating a chain of events between closely connected social start-ups. The ultimate goal for rural revitalization is to build an economic cycle in each area that facilitates the continuous development of social start-ups. While not all projects will or must reach this ultimate goal, the initiatives below are essential if you wish to aim toward it.

- Begin by enhancing diversity in the region and keep generating enthusiasm for change.
- Look for GPs, describe your area theory, search for your key business, and set up a digital platform.
- Develop new businesses through a chain reaction of events.
- Use data to describe the location and quantitative data to show the impact of your activities. As you do so, aim to slowly and gently create a large community that can aim for a common town development goal.

When tackling rural revitalization initiatives, it is important not to forget this overarching vision. However, the reality in front of you is that of modest adjustments and modest problem solving. Rural revitalization initiatives simply amount to the accumulation of consistent efforts at coordination and implementation. In any case, they will not succeed merely with one person, specific approaches, and set values. In reality, there will be significant gaps between the approaches and ideas of those involved with the project. When you set about engaging in a rural revitalization project, setting up individual meetings and choosing who to invite and when to meet with them can be a challenge. Moreover, expecting participants to keep the abstract idea of your project in mind is also a tough ask. But again, this is the reality.

A key factor here is the strength and cohesion of your team. You need people who can keep in step with each other. People who are good at communicating, good at putting together business strategies, and good at offering their expertise and ideas. People who have the stamina and courage to drive processes forward. The challenge is to have people who are interested in tackling the region's issues gradually combine their expertise and strengths, and to facilitate and develop this synergy. In the end, a good team is paramount.

The initial launch stage will be particularly tough. In this chapter, we have suggested putting together a framework to dictate the course of action for the project. We hope you can use this information as you assess the current stage of your project so that you can consider what is required in your area and how to assemble the right team.

Chapter 3

Interviews with the Trailblazers

Bringing our perspective of the region into focus as we move into the next stage —Mr. Hima Furuta

——The Trailblazers Conference has now come to an end with the final session 7. Do you mind sharing your thoughts and feelings?

First and foremost, I find the very structure of this conference body quite novel, and think it's an interesting endeavor. A small group of people come together and take turns presenting their own case studies, one at a time, which is then followed by an in-depth discussion open to everyone. It didn't sound like something that was feasible or possible to me.

These events are usually held in the style of a seminar or symposium, where one ends up listening to a relatively one-sided talk by an expert or a person who has achieved results or success in the field of regional revitalization. Even if we label it as a "discussion," it is difficult for everyone to discuss things if a lot of people prefer or choose to listen instead. Having said that, a simple conference or meeting is usually inadequate for anyone to delve deep enough into the issue. There are many events where various guests are invited to present or talk about their case studies. Even though they are able to share the information with everyone, I think it's not often that even the speakers themselves can gain new insights.

But this time around, not only were the members asked to introduce their case studies, but an overarching theme was also set beforehand for each session, with the sessions arranged like "chapters" so that each presenter could follow on what the others had done for rural revitalization. Take, for instance, how I talked about going from independent financing to mutual-aid financing in session 1, and how Mr. Takemoto talked about going from public aid to mutual-aid financing in session 2. So as we continued on with our discussions, it felt like our perspectives and ideas gradually came into focus and afforded us the ability to delve deeper. And I think that was just fantastic.

I also liked being able to watch the videos or interview footage about the case study before each session started, as well as at key points during the conference. This is all thanks to Mr. Hori and Ms. Miyase, who covered the events, interviewed the people in person, and compiled them for us.

——What kind of effects do you think the videos have?

I think it's extremely important that the videos were taken by a third-party journalist from an objective angle. It is not unusual to see videos used in presentations—my team and I have made many videos introducing our projects. But those videos were made from our perspective.

However, Mr. Hori and Ms. Miyase did it differently. They looked at things from a different angle and asked questions like, "What is this?" or "How is this going to help with that?" And I think through these questions, they were able to draw out the important elements that we could not properly put into words. And because they took the same approach of interviewing the people and covering the events in person for all seven case studies, we were able to follow their path and view things through the same lens, which made the stories and discussions feel connected. This had a synergistic effect that further clarified our perspectives on rural revitalization with each discussion session, and got us thinking more deeply than ever.

As facilitators of the conference, Mr. Hori and Ms. Miyase also interjected appropriately with questions like, "Is this what you mean?" to break down the main talking points of the discussion from an objective standpoint. I think if it had been just us and Mr. Murakami, the discussion would have drifted more and more in a technical direction, and it would have been difficult for outsiders to understand.

——Does that mean that the conference went almost as planned?

No. Even though the discussion was "chaptered," it led us through to the end gradually, and I think no one actually knew from the start where the conference would lead us. It felt like we were taking a journey over the mountains, and we hiked through one mountain pass to see the next one come into view.

To put it another way, we climbed up the same mountain via different routes and got to meet each other when we reached the summit, saying things like, "Oh, I didn't know there was a route like that," or "I'd like to try that route next time," as we shared information, accrued knowledge, and envisioned the next mountain to conquer.

In that sense, it was good that we were able to take our time to get to know and understand each other as we proceeded with the discussions. It also felt like we were able to gain knowledge together, too.

——I guess that means that the unexpressed experiences we all shared were explicitly put into words. Were there any new discoveries or revelations for you?

Yeah, for me, that would be the case of Ama I mentioned earlier. I thought the approach of using public aid to try to revive the town's high school, which gradually transformed into a framework of mutual-aid financing as more people got involved, was amazing. There are different types of mutual-aid financing as well—there is mutual-aid financing that is similar to independent financing, and there is mutual-aid financing that is similar to public aid. And I think there are various ways to seek mutual-aid financing accordingly.

I also learned a lot from Mr. Maki's case in Nishiawakura, like how to start small and steadily expand a project, and how to roll it out or what to do after it has reached a certain scale. This is how we move on to the expansion stage, as Mr. Murakami calls it. Precisely because all of us have been at that stage in some way at one point or another, there was a great deal that we could relate to. I was also very inspired by the way Mr. Maki's company is seriously developing human resources, taking employment into consideration.

——After this conference, what would you like to do next?

I think that we could visit actual places intended for rural revitalization and hold similar conferences in each place so that the locals can see for themselves what is going on. Doing this in Tokyo doesn't help them relate or connect to it, whereas doing it locally would give the issue relevance to them. And I also think that it's easier to pull them into the momentum if all of them are in it together.

The other thing that I'd like to try is to get large corporations involved as well. If possible, we are hoping to get people at the management level with some clout to participate and deepen their awareness of rural revitalization. There are quite a few corporate folks who oversee or lead regional development study groups, but when those people want to get their organizations to take action, they still need the understanding of top management.

Ms. Fujisawa has talked a fair bit about investment and social impact at this conference, and I believe that there are many ways for companies to get involved in rural revitalization, not in the form of so-called CSR (corporate social responsibility), but in a way that can be profitable for them as well. And I look forward to such developments.

Exploring the components of rural revitalization, from proper names to functions —Mr. Yoshiteru Takemoto

——*What did the Trailblazers Conference mean to you, Mr. Takemoto?*

It got me to delve deeper into my thinking, and allowed me to see the elements and functions of what we have done up to this point in a new light, and reorganize them.

For instance, Mr. Abe used diagrams and figures to give a lovely explanation of the initiatives he has been driving with the Echigo-Tsumari Art Triennale, and I felt like what he said really lit up my brain. Which parts of the entire initiative or project were Ridilover responsible for? Which parts do they need to work on more, and what are the challenges they would need to address down the road? This project also gave me a very good understanding of what Mr. Abe meant by accepters, committers, and influencers.

At the same time, Ridilover succeeded in elevating the roles of those who first started working on revitalizing the region—that would be Fram Kitagawa and Mitsu Hara, for the Echigo-Tsumari Art Triennale—to one of "functions." And I found that extremely inspiring and interesting.

——***I think you have already incorporated these elements and functions in Tobimushi's activities, am I right?***

Yes, I would think so! Whether it was myself, my company Tobimushi, Inc., or the people who are responsible for various functions in the company, it was helpful to be able to reorganize their roles by trying to apply these functions to each of them.

The term "senary industry" often comes up in regional revitalization, referring to the integration of production (the primary industry), processing (the secondary industry) and sales (the tertiary industry) (thus, 1 x 2 x 3 = 6, from the Latin *senarius*, "consisting of six") by using local resources. But I believe that going from the primary industry to the senary industry itself is not the goal, but rather the inevitable result of transforming regional businesses into sustainable ones. Likewise, our activities involve venturing into unfamiliar places, starting a project or business while interacting with locals in various ways, and creating a market that can only be found there. This process, I believe, inevitably requires certain elements and functions, and this is what I have been able to see in a new light.

——The initiatives developed in Ama centered around the reform of the high school. Could "changing the region with education as the springboard" be a template for rural revitalization?

I think the word "template" carries with it a slightly different nuance, but when it comes to school education, basic municipalities like cities, towns, and villages can only exercise jurisdiction over compulsory education, while high schools are under the jurisdiction of prefectures. This is why basic municipalities may not be able to take the lead to drive reforms involving high schools. This is in fact a bottleneck; it's why other municipalities that have seen and lauded Ama's initiative tend to think that it would be difficult for them to do the same thing.

The reason this initiative could succeed in Ama is because it is an island, and if the only high school on the island closes, young people will have no choice but to leave. Municipalities on land are in a different situation. Ama wanted to prevent its young people from leaving the island, even if the town had to use financial resources originally intended for other purposes to make it happen.

In other words, each region has its own situation and characteristics, and it is difficult to just use a successful case as a model and apply it to the region without any change. To go further, it's not that we can narrow it down to five winning models after comparing and studying a number of successful cases. Rather, we look at one region to try to get a feel for whether we could possibly apply methods that worked well in other similar environments or incorporate elements that seem to have brought about results in said region. Or perhaps we see that there is a need for a common understanding, perceptions, and attitudes. Those are the nuances I'm talking about.

——I guess the same can be said about individuals serving specific functions.

Yes; individuals as functions, not specific people. For example, many people think that Mitoyo was able to succeed because of Hima Furuta, the genius mastermind behind it. And that is true, but theoretically speaking, it should be possible for other people to take similar actions by breaking down the functions of Mr. Furuta into their component elements. And if we look at Mr. Kado, the responsibilities that he has had to carry alone could potentially be split and shared among several people. I would say that one objective of this conference was to analyze these elements.

However, that doesn't mean that we are going to dismiss the idea of relying on people. What I'm saying is that even though we don't necessarily need a Mr. X, we still need other people who can perform the functions that Mr. X served. In that sense, it might be better for us to say that when it comes to rural revitalization, we should not be looking at proper names (Mr. X), but rather specific people who can *function* as Mr. X.

In any case, I think that the participants who have gathered here at the Trailblazers Conference have all been expanding their activities as individuals under their own proper names, and there is a strong desire to move away from that and break individuals down into functions.

——Would the next step be to increase the number of people who are capable of carrying out such functions?

That's also part of it; it is true that we need people to revitalize the rural regions. I believe that what we need are not proper names like a Mr. A or a Group B, but independent-minded individuals who can collaborate with each other and fulfill their roles.

Earlier, I talked about how this Trailblazers Conference has helped me to reorganize the various things we've done for rural revitalization, but I still feel that we did not really go into detail about the positioning and roles of people like GPs, area organizers, and accelerators, as Mr. Murakami mentioned. The fact that we have realized that there is still a lot of room to explore this issue is an achievement in itself, and I hope that we will be able to discuss this further in the next phase.

Leveraging the power of foreign media to create a profitable region —Mr. Yasuhiro Kamiyama

——*You talked about lobbying activities in your presentation at the Trailblazers Conference, and I think it was quite inspiring for the other members.*

It's true that what I was doing may have been slightly different in nature than what you guys are doing. As Mr. Maki said, I think it's great to have a "bonfire" approach, where you go into a community, bring sparks of business ideas to life, and let the enthusiasm gradually build the small fires into a massive one. I used a different approach in my case, where I worked with the government and politicians, and I hope I succeeded in communicating that there's this way of climbing the mountain as well.

But I also learned a lot from this Trailblazers Conference, that each region has its own set of circumstances and environment, different ways of doing things, and everyone has their own way of thinking as well. It was inspiring and fun. The discussion allowed me to visualize conceptually what I had felt personally and concretely by talking with everyone here. It feels like I can use the abstract concepts or ideas that formed in my mind and apply them in a concrete way back in my own case or region again. I guess you could say that whatever was conceptualized was put into words and reorganized.

——*Do you feel that your actions or thinking have changed, then?*

The circle that I am in now is basically tourism, and I'm beginning to think that it might be possible to add a GP-esque function, as we discussed at the Trailblazers Conference. In a broad sense, the role of the GP is about finance and financing. But how can I incorporate that?

Take the domestic tourism market, for instance. Since it is declining overall, we could bring in foreign capital for investments or ownerships, but let the community or locals take over the operations. If we were to just sell the land and stop there, it would do nothing for the community. Instead, we could create a fund, start a business, get GPs as well as various LPs involved, and let the community manage the business after it becomes profitable. And I think many of you talked about that at the conference.

——So your idea is to nurture sparks of initiatives throughout the community while getting the locals to retain ownership.

Yes, you could put it that way. And for that to happen, I would have to become better at creating organizations, as well as getting the locals to cooperate and collaborate with each other.

After all, it's not going to work if we can't create a system or framework that keeps the local economy going. We have to make money. The kanji "儲" (profit) is literally made up of characters that together mean "信者" (believers), implying that initiatives or projects will never become profitable if the number of believers does not increase. Since there are bound to be a few motivated people in every community or region, we can bring them in and connect them with those in public administration. I think we should actively seek involvement not only from the private sector, but also from public administrators and politicians, because after all's said and done, they have a great influence on the community.

——Speaking of foreign capital, you also mentioned that people find praise of their region more convincing when it comes from foreigners.

Yes! When they are praised by a Japanese, it somehow comes across as flattery or polite niceties, but sincere praise from foreigners makes them genuinely happy and in turn bolsters their confidence in their community or region. Isn't that another way to achieve revitalization? In this sense as well, inbound tourists are important.

This applies not only to tourism, but to sports exchange as well. An example would be the Town Revitalization with Baseball project in the city of Anan in Tokushima Prefecture. Instead of professional baseball teams, the initiative focuses on bringing in various amateur baseball teams from all over the country to compete and entertain the players. We could do this for other sports as well, like rugby and soccer exchange games with teams from foreign countries.

Another idea that comes to my mind now is to hold many small-scale MICEs (meetings, incentive travel, conferences, and exhibitions). A few years ago, Aomori Prefecture, which is famous for its garlic production, created an NPO called the Black Garlic International Conference. They started holding the International Black Garlic Summit on a yearly basis, and now black garlic ambassadors from the UK and France attend. I think such initiatives are another way to revitalize the region.

The point is to create reasons for people to visit the region. Creating demand is really important. And this is also where the value of public relations comes into play.

——*So how should we communicate the value of public relations?*

Many are now saying that it is important to use social media to effectively spread the word. I won't deny that, and I think it's a good idea to make good use of social media, but to me, it is the voices of journalists that have the most impact on return on investment. They are skilled in accurately capturing the essence and insights and expressing them in written articles, photos, and videos. Communication by a professional who possesses such media techniques can have a huge effect, and can also be the source of material posted on social media.

Foreign journalists are especially good for this. If possible, I'd like to organize a media tour that invites journalists from abroad to visit Japan and report on what they really think and feel about the regions. The "Castle Stay" in Hirado that I'm involved with was featured on the BBC, which brought in applications from European tourists. Recently, Morioka was selected as one of the "52 Places to Go in 2023" by the *New York Times*. The city responded to this with surprise and bewilderment, although they hastily started putting in efforts to promote their tourism. It is also critical to make good use of such external influences.

I think this Trailblazers Conference has also made me realize, once again, the importance of the power of communication, and I'd like to continue to create demand by sticking to my own basics of how to create local content.

Bringing rural revitalization to the highest levels with "investment theory" —Ms. Kumi Fujisawa

——*What has this Trailblazers Conference meant to you, Ms. Fujisawa?*

What I took away is that rural revitalization can be generalized in two ways, and I think that is extremely meaningful.

The first is the generalization of phenomena. This also applies to entrepreneurial success stories, but many success stories of rural revitalization often only give a factual report on the things that have been done, and there is already a lot of such information out there. Despite this, many have found that rural successes are not that easy to duplicate.

I believe that this conference has elevated the various elements involved in these phenomena to a higher level. In this conference we have seen illustrations and diagrams of these elements that have also been put into words, so when another person tries to do the same thing, they'll know what to do and where to focus.

The second is the generalization of emotions. The invaluable video footage created by Mr. Hori and Ms. Miyase objectively conveys the emotional aspects of the hardships, anxieties, and expectations of those working in the regions and communities from a third-party perspective. These emotional aspects are usually not put into words, but it was wonderful to be able to see and witness them through the power of video.

——***This Trailblazers Conference has brought together members, including you, who have been able to talk concretely about what they have done, as well as their abstract concepts or ideas. How do you feel about this?***

I think that is exactly why the conference was able to generalize rural revitalization in this way. At first, though, I wondered if it was really okay for me to be here. Each and every member here is very capable of driving actions on the ground as well as thinking deeply. But personally, I still feel that something is lacking in me when it comes to working with others on the ground, and that is why I was so inspired by what you all said. It made me feel that I have to do more, too.

It was actually around the time of this conference that I was invited to join the Institute for International Socio-Economic Studies (IISE), the think tank where I work now. And I eventually made the decision to move there because I thought that I could get a real feel for what it would be like to work on the ground. So I would say that the Trailblazers Conference had a tremendous impact on me, even in my career decisions.

Since my work up to that point had mostly revolved around giving advice, I was drawn to the idea of working on the ground. Thanks to all of you, I am now able to do work that has a real-world feel to it in many ways, and I'm really putting in my sweat equity. [Laughs]

Aside from that, every trailblazer member is a completely different character who has done amazing things and has very interesting stories to tell. In the beginning, each member had a commanding presence, like a distant mountain, and we were waiting with a little awe to see what they would say. But by the end of the conference, it felt like all of us could relate to each other on the same level, as if all the mountain peaks or the underground waterways were connected.

——Are there any stories from the conference that particularly impressed you?

Rather than a particular case study, I was drawn more to the human aspects of everyone's stories that you just don't find in textbooks. When we talk about projects, the focus is often on methods, such as how to raise money or what steps to take to make it work, and not so much on the difficulties or resourcefulness involved in bringing people together toward a single goal. I thought it was great that we were able to get a glimpse into these aspects through this conference.

——Your suggestion about applying investment theory to rural revitalization has triggered a discussion on how to create a kind of template for this work.

Throughout this conference, I was repeatedly reminded of the importance of finance or financing systems. If you are driving a rural revitalization project on the ground, but the scale is still somewhat lacking at the final step despite having achieved sizable results, then I'd say that one of the impeding factors is how the money was used.

The most fundamental part of getting money is, without a doubt, disclosure, which is to effectively convey the outcome of the project to the investors or donors. This is not just limited to money; it also serves as the basis of management, like how we measure the extent of a person's growth if we invest in them.

It then occurred to me that perhaps the future challenge for rural revitalization is to incorporate more of that management perspective and bring the connection between social impact and business closer. For someone like me who specializes in entrepreneurship, the hurdles are not so high if the only issue is money. There is already plenty of knowledge about how to revitalize the community, and if you add investment expertise to that, I think we can achieve the highest level of rural revitalization.

And I think that this is actually not just a regional problem, but one that traditional Japanese companies, or even Japan as a country is also facing, in that something is keeping them from achieving significant growth at the moment.

If this Trailblazers Conference is to continue in the future, I would like to discuss how we can strengthen the financing and investment aspects, and how we can increase the number of people who have the expertise to support these kinds of revitalization projects.

The need for true "pacesetters" who will not be swayed by the priorities of supporters —Mr. Shintaro Kado

——*What are your thoughts and feelings after attending the Trailblazers Conference?*

Even if I were asked to sum it up, rural revitalization itself is a massive, overarching theme that includes a diverse range of methods and approaches to the projects or initiatives, depending on the region. I think it was amazing that we were able to see the various approaches used in rural revitalization and that we were able to express those elements in a common language.

I also think that the inclusion of perspectives such as finance and the role of general partners was meaningful in changing the image of rural revitalization that has been talked about thus far, or rather, in shedding light on other aspects of driving rural revitalization.

The function of a GP in a finance context is slightly different from the function of a GP as discussed at this conference. But for us to move forward from a management perspective in rural revitalization down the road, we definitely need a GP-like person to proactively manage the project or business and "accelerator"-esque roles to support it. So it was great that our discussion included the need for a "kelp forest," a safe environment to nurture such people.

On the other hand, many elements were discussed from various aspects and angles, and I think it is important to combine those parts in a skillful way to create a story. The ultimate aim is to create start-up ecosystems for rural revitalization that can be applied to any community or region. And I think we have come to a common understanding that this is essential.

——***Were there any cases or discussions that particularly left an impression on you?***

I would say the initiatives conducted in Ama and Nishiawakura, in the sense that I could relate to them. Like how they approached the business or project entities, how they came up with plans, how they got various parties or stakeholders to be involved, and how they came up the frameworks and venues for them to serve their functions properly. I thought the whole thing was amazing, and felt that these projects could very well serve as a model for others.

When Mr. Takemoto talked about Ama, he said, "In order for one to survive, everyone must survive." This struck a deep chord in my heart, as I have gone through many hardships in my own community. Because there are people from all walks of life with all points of view, we simply cannot move forward unless we have the management skills to convince even the most stubborn opponents to change directions. I also thought that it was wonderful that the architecture, which included these aspects, would undoubtedly pay off.

——In your case, you were a party concerned with the local business entities, and you also played the role of a supporter from outside the region. Do you have any thoughts from that standpoint?

How can we best reconcile the incentives of those who are passionately engaged in a project within the community with the incentives of those who are investing in the activities from the outside? It seems to me that it would be important to take this perspective into account as well.

I was under the impression that this conference would focus on the priorities of those who provide support to the community or region. And that was, of course, meaningful, because it provided a language and a definition of the elements and functions necessary to increase the population and funding for rural revitalization.

But having said that, even if the people who provide support are encouraged by this, I am not so sure that those who do the backbreaking work on the ground in the community will be motivated to go for it or do their best after hearing this. Considering my own experience of struggling on the business side, I am very curious to know what motivates people to take the initiative and follow through with a business or project, even if they have to keep deferring to those around them. Why take the risk of launching a business project on your own? And why are they so passionate about community building? I hope that one day, we will be able to come up with an outline or framework that will be convincing to the local community, including these kinds of perspectives.

For that reason, I think it would be good to have a place where not only the trailblazers who have had successes can gather, but also people who are struggling alone on their own on the ground, or people who have failed, and let them talk from that perspective.

——Mr. Kado, what kind of developments would you like to see happen next for rural revitalization to move forward?

If we're talking about the future of the Trailblazers Conference, how about having, for example, members of the Trailblazers Conference to go to the site of an actual initiative or project in a region and giving them free rein? They would be asked to think about the social impact, create a fund, and also put in place GPs and accelerators. They would also have to consider the incentives for those in the community who are putting in a lot of effort. I feel like this would gradually bring the elements and the know-how required by the pacesetters who support local initiatives into better focus.

Ultimately, rural revitalization is about the creation of start-up ecosystems that function on their own even without amazing trailblazers. And for us to do this, we must incorporate social-impact financing and have pacesetters to support the initiatives and businesses. I believe more discussion is required on how we can go about creating such a framework.

Personally, I would like to promote the creation of a market based on mutual-aid financing. The current reality in rural areas is that public aid alone is not enough to support any kind of revitalization, and independent financing alone is not enough to keep it going. I believe the only way to turn this situation around is to expand the framework of mutual-aid financing, and for this to happen, the use of digital platforms and open data is indispensable. The success or failure of rural revitalization will ultimately depend on it.

Creating a framework to systematically combine the assets of connected minds —Mr. Toshiki Abe

——*What did you think of the Trailblazers Conference? Was there anything that you found interesting?*

It was very engaging, right from session 1. The case study of Udon House was extremely interesting! Drawing foreign inbound tourists with an udon noodle restaurant where they can stay for 30,000 yen a night, and then turning it into a profitable business. And with that successful initiative-turned-business, they further expanded it, went on to get the locals involved, and started various projects one after another. In theory, it should work, but in reality, there are very few cases in which this has been put into practice so beautifully. It feels like Hima Furuta was directing an ensemble cast of young people who were trying to weave a story on the ground in the community as they dealt with conflict—it was simply mind-blowing.

I never had any previous contact with Mr. Furuta, but after talking to him this time, I felt like he really knows how rural revitalization works. And as the conference progressed and covered more and more topics, we talked about things like GPs and finances, but Mr. Furuta was able to wrap up the conversation succinctly and precisely. I think it is really quite rare to find someone who has such a keen on-the-ground sense as well as financial literacy.

Of course, all the other members were amazing as well. There are very few people that I can talk about this with in-depth and who understand me, not to mention helping me gain a better understanding of myself and what I'm doing. I personally really enjoyed this Trailblazers Conference; it gave me the chance to talk with these people on a regular basis. Overall, I would say that the format of the conference was very well done.

——***I understand that you have known Mr. Hori for some time. What did you think of the video footage?***

I think the video footage was nicely done. I'm glad that Mr. Hori did not get too caught up in the voice and tone, and showed the actual situations as they were. Actually, Mr. Hori and Ms. Miyase were amazing. Despite the fact that they themselves were not very familiar with rural revitalization at the beginning, they seemed to grasp the essence of it with each session and case study they worked on, which is reflected in the changes they made to the way they conducted interviews and captured footage. I found that transformative process interesting as well.

Our goal at Ridilover is not so much about the on-site operations at the micro level, but more about driving many different initiatives to create a macro-layer framework in the background to lay the foundation, and this is very hard to capture through video footage to begin with. I feel Mr. Hori and his team did an excellent job. Their videos perfectly complemented and refined the image of what is actually being done in the communities.

——***Some of those working on the ground have said that being interviewed has increased their sense of involvement.***

For sure! Being interviewed is kind of like the study tours we conduct, with outsiders coming and interacting with each other on the ground. But there is a difference. Great changes occur in outsiders during those study tours, whereas the reverse is true for interviews—they make great changes occur in those who work in the community or region. And this is how people become committed and dedicated to what they are doing.

That's what we are good at—creating these kinds of opportunities. At this conference, I have set out to explain the steps involved in creating an ecosystem for addressing and resolving social issues. And I don't think there are any other players who specialize in this kind of thing. If the role—or rather, the necessity—of this conference had been properly communicated to the trailblazers, I think it would have been even more meaningful for Ridilover.

That being said, it is one thing to understand the importance of what is talked about, and another to be able to actually reproduce it. It is exceptionally difficult for someone to create a business structure that also forms a safe environment to nurture people engaged in rural revitalization—what we call a "kelp forest"—so it is important to first let people know about such efforts and spread the word.

—I guess one could say that Ridilover, as an organization itself, could very well serve as an ecosystem. What are your plans for the future?

That is certainly one of our goals, and we're working hard to "combine the assets of connected minds," as we say. Newcomers face particularly tough requirements in working on the ground and dealing with social issues or running community activities. I'd say that the expectations for newcomers are unusually high. This is why a lot of people enter our field with great enthusiasm, only to end up sadly giving up in the face of the overly high walls. Ridilover serves as a place to kind of shelter them from that and help them to grow by accruing experiential knowledge.

Our employees are encouraged to visit the communities and work on the ground so that they can experience everything for themselves. What they accumulate through their many adversities and hardships becomes not only their personal experiential knowledge, but also the organizational memory. And as the organizational memory grows, when other members come across a similar situation in another area, they will be able to easily and smoothly get through it.

Until now, this was mostly done within the Ridilover organization, but I believe that from now on, we must work together with people outside of the organization to develop a workforce that can achieve rural revitalization.

Using the rural revitalization "Trailblazing Model" to power business expansion —Mr. Daisuke Maki

——Has attending this Trailblazers Conference clarified anything for you?

It really feels like this conference has provided a common language that has helped give shape to things or ideas that were previously fuzzy, or at other times, nonexistent, in my head. I think it has served as an excellent process that allowed us to organize and articulate the important concepts we need to think about the future of the rural communities.

I myself am usually buried in on-site activities, so time to think about things systematically and deeply like this is very valuable to me. I'm therefore very grateful to be able to attend such a conference on a regular basis. Besides, I also think it's easier for me to formulate a tentative theory of what to do once a framework has been established in my mind, instead of just stopping it at the articulation level.

In fact, right now, our company is starting something new in Kinko, a town in Kagoshima Prefecture, and I have a feeling that what we have conceptualized at this Trailblazers Conference will be useful. Because every region presents a different set of circumstances, there are some aspects or areas where experience gained in one region alone is not enough for us to skillfully deal with challenges in another. In that sense as well, I'm really thankful that I was able to learn specifics from other regions and extrapolate them into abstract concepts at the same time.

——I suppose it is important to know what's happening on the ground as well as the theory.

Yes, you could put it that way. It is important to have a keen on-the-ground sense to begin with, but you can't possibly have that until you are there in person. Without the video footage shot by Mr. Hori and his team, the conference would have seemed clinical and detached to a third party, or it might have just ended up being a bunch of experts having a somewhat difficult discussion in Tokyo. Despite the fact that we could not actually be on-site in person, we were able to share the voices, expressions, and atmosphere of the people who live and breathe in those communities through the video footage. I think that's why we were able to expand the discussion. In that vein, I would like everyone engaged in rural revitalization to know what's happening on the ground.

As Mr. Abe mentioned, it is meaningful for the local people to have the opinions of a third party, too. Visualizing the narratives and accounts of the land in the form of footage will allow us to have that "aha" moment, "Oh, that's what it's all about!" Even though Nishiawakura is just a small village with a population of less than 1,400, every one of those locals has their own story to tell. Not all the stories are shared by the villagers, so being able to draw some context from it is by itself a discovery to boot.

I mentioned this at the beginning, but if we only talked to our coworkers or stayed within our circle in the community, we would almost never have a chance to look at things in an abstract way. But if we don't do that at the important junctions, it will be difficult to reproduce similar initiatives elsewhere or to move on to the next stage. This is one of the things that I took away from this conference.

——Based on this experience, Mr. Maki, do you expect to see any changes in your future activities?

The first thing I want to work on is how to set up the warm-up stage in Kinko, which I mentioned earlier. I feel that just being aware of this initial stage can make a huge difference in the quality of our work and the way we carry it out.

As a matter of fact, we are going to visit Mitoyo in the near future, and I plan to stay overnight with my employees. I want to learn more about the basic infrastructure model that Mr. Furuta spoke about at the conference by talking to him in person right where it was created, as well as gain a firsthand sense of what we've never experienced or seen before. I think this comparison will help us to make new discoveries and also better understand the characteristics of our own region.

The inspirations that I got from the Trailblazers Conference—I do not want them to be limited to me. When I told my employees about it, they were all very eager to learn more, and eventually, 10 of them decided to visit Mitoyo with me. You could say that this kind of interregional exchange is one of the outcomes of the conference.

One change that has already occurred is that we are now looking at income as a key performance indicator (KPI) for our business. I haven't paid much attention to this in the past, but I now know that if I want to scale up my business down the road, I will have to set "increase in regional income" as one of the KPIs. At A-Zero, our Local Venture Development Division has quickly become aware of the taxable income and number of taxpayers in the regions where we are working.

——*I guess your next goal is to advance to the expansion stage.*

We would like to expand the scale of our business and invest in various prospects that hold economic potential, as well as turn rural regions into places that can attract investment. But the inherent nature of our work means that instead of just going straight for an IPO, it's really more a matter of building up our experiences and achievements bit by bit. And honestly speaking, we are still at the stage where we don't really have a roadmap in sight. We need to eventually figure out our company's capital policy, as well as that of the regions, to see what types of funding options are possible. If this Trailblazers Conference continues in the future and we can have discussions to explore different approaches to the expansion stage, I would love to participate again.

Putting into words the unspoken thoughts and voiceless voices hidden in the community —Mr. Jun Hori and Ms. Mayuko Miyase

——I understand that both of you went on-site to cover all seven sessions of the Trailblazers Conference. What kind of footage did you set out to capture?

Hori: I was from NHK, while Ms. Miyase was from the Fuji TV network. Because both of us have been in the world of news reporting for a long time, we know that there are still many things in the world that have yet to be put into words. Our desire is to articulate abstract concepts and human subtleties that are difficult to convey through mere words or photographs, and the true feelings that even the person speaking is not aware of. That was what we had in mind when we went independent and established the Watashi o Kotoba ni Suru Research Institute.

So when we were asked to collect information, conduct interviews, and capture video footage for the Trailblazers Conference, our first thought was to create a format that would allow the seven trailblazers, who would be presenting their successful case studies of rural revitalization, to make new findings or have new revelations. Throughout the process of capturing footage and creating videos, we conducted various interviews with the trailblazers and with the locals in the various communities, and prompted them to express their unspoken thoughts and feelings; these interviews would serve as the basis for the workshops. We also felt that such workshops might provide an opportunity for all of you, who are living and working in the rural regions, to gain a fresh perspective on your daily lives and projects, and help open the next door of opportunity.

In typical news reporting, there is a limit to the number of places we can visit and the time available for covering a theme or topic, so it is difficult to get right to the heart of the matter and reveal the true feelings and honest thoughts of those who are interviewed. That is why, at this Trailblazers Conference, we went so far as to ask all of you to talk about the things that were difficult to discuss, on the premise that they would remain behind the closed doors of this conference. On top of conducting advance interviews and covering the activities on the ground, we also decided to videotape your discussions at the conference so all of you can evaluate them with an objective eye.

Miyase: We invested a lot of time in these preparations. Before visiting a region in person, we talked with the related trailblazer at least twice. Based on these interviews, we then came up with a tentative concept, decided on the keywords, formulated a general framework, and worked out who we should meet and talk to in the community. Overall, I'd say that it took us over a month to prepare for each session.

Having said that, every time we visited the rural regions to collect information and conduct interviews, it was a mad rush, since all our trips were limited to just one day due to the budget. [Laughs] We'd be in a rural region for about five hours to interview the people we had arranged to meet in advance, but only about half of the people we met were candid with us. The rest of the footage consisted of people we could only meet there, accounts and narratives that no one would or dared to tell us, and things we didn't see. This meant that we captured an immense amount of footage, brought it back with us to Tokyo, and then edited it down as well as we could to just 15 minutes.

Hori: We really had to roll up our sleeves and buckle down, as it was a grueling task. Every single community is filled with stories that could fill up a book, and our job was to extract the essence from them.

—Even though you promised that whatever they said would stay within closed doors, wasn't it still quite difficult to elicit the honest thoughts of the community?

Hori: I think there are two kinds of honest thoughts; one is about facts that have not been disclosed before, and the other is about things in a person's subconsciousness. For this conference, the latter was particularly important. There were many times when they were able to put their honest thoughts into words and express their feelings for the very first time because someone—that is, us—from outside of their community went to talk to them, and listened to what they had to say.

Take the Echigo-Tsumari Art Triennale presented by Mr. Abe, for instance. Ridilover has been organizing study tours and helping to develop a workforce. And after we talked to Mr. Abe and other members, it occurred to us that the structure was akin to a kelp forest in the ocean.

In fact, when I first started out as a journalist, I was working for a broadcasting station in the Seto Inland Sea, so I knew about the existence of kelp forests, which provide a safe environment where fish can grow up undisturbed, and are essential to the sustainability of the marine environment. The concept overlapped perfectly with what Ridilover was doing. Based on the keyword "kelp forest," we then came up with a format for the conference: Finding out how the rural revitalization project came about, its function, the people that came together to work on it, and what happened next . . . and then weaving these parts into one context and narrative.

Another thing that I want to mention is that as journalists, we decided not to just make introductory videos. We kept that in mind when we were depicting the case studies as well as the hidden dilemmas and quandaries in the communities. We also did not want to portray successful case studies as mere success stories. We wanted the video footage to honestly reflect what we'd seen and felt from our visits and interviews, including the hardships, friction, and even contradictions, in the hopes that we could uncover things in their subconsciousness. That is how we approached this project.

The voices of anonymous citizens are sometimes left out of the process of designing a social system, you know. They only get media coverage when they become a social problem; otherwise, they are often overlooked. I think the meaning of our involvement in this project lies in drawing out the voices of ordinary citizens, not because they were government officials or activists, but because they were simply there.

—Is there anything that you have noticed or changed since conducting the interviews?

Miyase: That success in each region is never just down to the efforts of one outstanding person. The objective of this conference, and our goal, is to explore forms of rural revitalization that do not center on any particular person. But even though I was well aware of this, a part of me still believed that the locals in the rural regions were, in reality, on the receiving end of the decisions made by the trailblazers. That changed when I visited them.

I realized how wrong I was when we arrived there. It was the same in every community or region. There were people from all walks of life who were engaged in rural revitalization of their own accord. Even though it was the trailblazers who kickstarted it, I could see how the locals had inherited the mission, passion, and capability, and were further driving revitalization in their own communities. They also were trying to find the next leaders amongst themselves.

It's not a top-down hierarchy, but connections that extend laterally. It was clear as day, and if I had to put a finger on it, I'd say that there are many talented people that are lying dormant like treasures—locals who are capable of bringing about change in urban areas. I believe discovering such people will become important in the future.

And now, whenever I get to see or hear such people, it makes me want to join the world of rural revitalization as well. I want to try to do something about it, too, and I'm actually pretty serious about it!

Hori: [Laughs] I was really encouraged and inspired, too! One of the things I learned was the importance of finance, disclosure, and governance. And when I thought hard about it, these things arose in the course of our reporting activities. When I observe how various incidents, accidents, and problems are handled, two questions always pop up in my head: Why was the information not disclosed? And who is to be held responsible? Looking at the trailblazers, I am once again reminded that if we design things properly, including the financing framework, things should improve. I think this is a common language that can be applied to all social issues, even if the fields and contexts are different. I feel that I now have a new vocabulary that I can apply in my activities from now on.

——***Which case study, would you say, left you with the deepest impression?***

Hori: For me, that would have to be Mr. Maki's project in Nishiawakura: how the media failed to see the real picture. I was at the Okayama Broadcasting Station when Nishiawakura made the decision not to merge with other municipalities, and I was on the side that reported their decision from a negative perspective. With depopulation continuing and even the survival of the village in jeopardy, I wondered if they had really made the right choice. Being, perhaps, at the mercy of the social climate, the trends of the times, and the words and deeds of the masses, the press set the tone by allowing themselves to be swayed by the phenomena at hand.

Fast forward to over a decade later, Nishiawakura has managed to become a trailblazing example of rural revitalization, as presented by Mr. Maki. Now that I have witnessed this, I am acutely aware that neither I nor the rest of the media at that time were able to see the true nature and potential of the region, and we did not have an eye for the calm verification of facts.

Times have changed, and now there is a movement toward "solutions journalism," in which the media tries to get involved in resolving the problems. The case of Nishiawakura has made us realize this new role for journalists. I also think the Trailblazers Conference has taught us the importance of a place for people to discuss and confirm our values—which change in response to circumstances—calmly and closely, with diverse knowledge and insights.

Miyase: There are so many memorable scenes that it's hard for me to choose just one, but I thought that Mr. Kamiyama's superhuman lobbying skills and activities were amazing. Mr. Kado's 13 years of tribulations were also riveting for me, to the extent that I wondered if community development was something that required that much commitment and dedication. It made me think that perhaps there are many places across Japan that are at an impasse as what to do with their communities or regions. If we are able to show people various ideas and methodologies, then I hope that this kind of conference will continue to spread the word widely in the future.

Personally, what impressed me were the keywords that emerged in each session. Various terms used by the trailblazers popped up in the discussions, which transcended the ones we had set during the preparation stage: "roles to play and places to belong," "encouragement," "spectators," "cell membrane," and "key business." As a person who works with words, I'd like to have the opportunity to spread these words further.

——What might you hope to see at a future Trailblazers Conference?

Hori: Now that I know there are many deeply secluded places that have yet to be seen, I would like to visit more of them. I would also like to see places that are working hard at building digital platforms. If we can manage to capture the hidden words, know-how, and dilemmas, I think the value of this conference will go up.

Miyase: This might turn into a somewhat long-term project, but what I would like to see is a group of trailblazers going into a community or region where nothing has been started or done yet, and start something from scratch. We would be very honored to be part of the making of such a verification documentary. Please leave it to us at the Watashi o Kotoba ni Suru Institute.

Conclusion: The Trailblazers Conference Has Taken the First Step to Being a Mutual Aid Society

Keisuke Murakami, Director-General, Digital Agency, Government of Japan
—EY Intelligence Platform Executive Office:
Mitsuhiro Sugata, Partner, Japan Region; Non-audit Services Markets Leader
Hiroyuki Nakata, Partner, Government and Infrastructure Unit
Hiroko Hashimoto, Manager, Government and Infrastructure Unit

First-ever "Trailblazers Conference" launched with a mission

——How do the members of the executive office feel now, looking back at the sessions of the Trailblazers Conference held over the past year and a half?

Murakami: My honest answer is that we started without really knowing what was going to happen. It doesn't sound very professional to say that we started with no clear prospects, but I recall feeling that we should just give it a go, because we had to take some kind of action.

Sugata: Yes, but our purpose was clear. Even in positive cases of rural revitalization which seem to have succeeded because of the existence of a specific leader, when you look closely, there should be a common model. We wanted to try to find a template for success from discussions among leaders whom we considered trailblazers in this field.

Nakata: Indeed. We put a lot of thought into choosing the trailblazers and the order in which they would give presentations at the conference.

Hashimoto: From selecting the trailblazers to deciding the layout of the venue, the schedule for each session, and the video interviews, it felt like we were creating an entirely new approach from scratch. The trailblazers are well-known figures in the world of rural revitalization, and I am somewhat amazed that we were able to bring them all together. I am deeply grateful that no one missed a single session.

Murakami: We had decided that Hima (Furuta) would be the first presenter. Starting with independent financing, then moving on to mutual aid is easy to understand, and Hima is a great speaker. After Hima, it seemed appropriate to hear from Mr. Takemoto, on the other end of the spectrum, who expanded the discussion from public aid to mutual aid. We struggled to decide who to ask next, but it turned out to be the right decision to ask Mr. Kamiyama to talk about his approach from an institutional perspective.

Sugata: The response to Mr. Kamiyama's explanation of his lobbying activities was powerful, and the atmosphere in the room quickly changed.

Murakami: The trailblazers all have similar experiences, but the lobbyist's experience is indeed rare. Out of the blue we were introduced to a completely new world, and I think many were surprised to hear that there is a way to make such things happen. This made everyone more relaxed and helped strengthen the bond between participants.

Nakata: At the beginning, everyone was waiting to see how things would pan out. My impression is that the discussions became more intense as common keywords emerged and thoughts were verbalized as more sessions were held.

Murakami: The turning point for logically organized discussion was the session where Ms. Fujisawa gave her presentation. After Mr. Kamiyama had warmed the room with his presentation in the session before, Ms. Fujisawa calmly covered the topic of finance, and I introduced my personal views on the structural theory of rural revitalization. I really wanted to include finance, disclosure, and governance somewhere, and it was in that session that I managed to ensure the smooth flow of discussion—the perfect time to start searching for a template.

Sugata: In the end, I remember thinking, "Ah, this is how a common language is born. It's possible to reach this kind of consensus." I was very impressed. In the future, we should be able to further develop the template and implement it in the field. I am looking forward to seeing what challenges lie ahead and what kinds of worlds we will see.

A group of experts who can move freely between the concrete and the abstract

——Where did the idea that it would be interesting to create this kind of body come from in the first place?

Sugata: About two years ago, when we launched the EY Intelligence Platform, we came up with three steps for action. (1) To create a place to derive a proper theory; (2) to formulate the theory and solidify it as know-how that can be put into practice; and (3) to actually put the theory into practice. The Intelligence Platform will work by repeating these three steps. That was the image I had in mind, and the Trailblazers Conference was the first phase.

Murakami: If the vertical axis is the initiatives being taken in each region, the horizontal axis is the common know-how that emerges from surveying such initiatives. In which case, a consultative body of wise people would be needed to look at the whole picture. The question was what kind of gathering this should be.

Sugata: Mr. Murakami had previously participated in a series of gatherings of astute experts in a different field who rarely get together. While the theme was completely different, he said that the experience was very enlightening.

Murakami: Mr. Hashimoto and I had connections to the trailblazers, and when we approached them to ask them to participate, they were intrigued and agreed to meet together.

Hashimoto: We put together a dream team of people that we were particularly interested in. I never thought it would actually happen, and on the day of the first session, I got goose bumps when I saw them sitting in a circle discussing the issues. I will never forget that moment.

Nakata: It is not unusual for consulting firms such as EY to hold meetings of experts, but as I mentioned earlier, they are usually biased toward either the vertical or horizontal axis, and there are virtually no opportunities for this kind of comprehensive, continuous discussion. That is why I was worried about not being able to see how things would pan out, but I gradually found my view of the world expanding, as if a fog was lifting.

Sugata: I have set up meetings between local governments and the private sector in the past, but we weren't able to broach theory and practice, or abstract and concrete issues, at this level.

Murakami: We were able to create a format that would not have been possible at a council meeting hosted by a government agency. First of all, budget-wise it is impossible for a government agency to make original videos for each session. [Laughs] Second, we were able to bring together these participants. I myself have been involved in rural revitalization for more than eight years, but it is rare to find people who can talk about the essence of the issues while moving back and forth freely between the concrete and the abstract. Third, we were able to create an opportunity for a class of people who are capable of giving keynote speeches on their own to sit in a circle and talk. As a result, I think they all found the experience interesting, too.

Sugata: One of the major achievements was putting together a meeting format that was neither government nor private-sector. In fact, we had a hard time internally at EY deciding whether to position the conference as a prior investment, branding, or policy recommendation meeting. We actually held the sessions without our colleagues fully understanding what we were doing, and so it is something of a relief to see such great results.

Murakami: They say that innovation always comes from the brink.

Nakata: I feel we finally have the wind behind us. According to a brand-image survey of key personnel at listed companies that we commissioned, EY Japan came out on top in the areas of rural revitalization and economic security.

Accelerated "empathy and understanding"; synergy while generating heat

——What stuck in your mind and what did you gain from the sessions?

Nakata: We heard interesting presentations at all of the sessions, and I was greatly inspired by the new keywords that popped up at each session. "Roles to play and places to belong," "half-civil servant, half-X," "nucleus and cell membrane," "magnets," "kelp forest," "spectators" . . . Mr. Abe's comment, "We don't have money, so give us money" was funny, and showed his true character. Personally, I feel that my own understanding gradually deepened as I worked on the minutes of the first and second sessions.

Murakami: After the conference ended, the trailblazers all concurred that things have come into focus. The feelings and theories that had been vague in each of their minds have now been articulated, and they have successfully come up with a common understanding. On top of that, the video footage was both effective and greatly appreciated. Mr. Hori and Ms. Miyase did a great job of filming objectively at each location, taking a vantage point somewhere between that of the locals and those who support them, and they were successful in getting both sides to say of the other, “Oh, that’s how they look at things.” I believe that it was thanks to their footage that a common language emerged naturally.

Nakata: The footage conveyed their high expectations for the initiative. The ensuing discussions quickly accelerated, based on a kind of empathy.

Murakami: In our theory of rural revitalization, we called this the warm-up stage, but the conference itself must first warm up the room. Just stating the theory is not enough. That said, relying on an emotional mechanism will not help us improve our understanding. Both are needed to get people motivated. The video footage did a good job of evoking our emotions.

Hashimoto: Just watching the footage at the last session made me cry. I felt really grateful when the trailblazers told me, “Don’t let this end here,” and urged us to move to the next stage.

Sugata: Some time after the last session, someone I had never met before told me, “EY seems serious about rural revitalization. I have never seen a consulting firm get so involved in rural revitalization.” We had published the minutes of each session in Mr. Abe’s *Ridilover Journal*, so perhaps that person read them and contacted me.

Hashimoto: After the last session, I was very happy to see all the trailblazers reunited at the symposium. After the symposium, someone approached me to say, “I have read all the minutes and watched all the digest videos. I was really looking forward to the symposium.” Another person who had flown in from far away just to attend the symposium told me, “Rural revitalization doesn’t always go the way we hope it will, and people around us sometimes wonder what we are doing, but today’s symposium made me realize that we are not doing the wrong thing.”

Shifting to long, slow-burning, bonfire-like activities

——How do you evaluate activities in terms of outcomes?

Sugata: In addition to what I have said so far, a major outcome for EY has been the name credibility we have gained. Of course, the Trailblazers Conference has reached the stage where we are starting to see a certain model for success, and while this model has yet to be implemented, we believe that it is a valuable step forward in demonstrating the significance of the conference both internally and externally. Creating this kind of discussion venue will contribute to Japanese society, where rural revitalization has become critical.

Hashimoto: As consultants, we are usually asking others to let us help them. Organizing the Trailblazers Conference has led to the new experience of being approached by people who have heard about the conference and want to do something with us. That shows how much has changed in the past year. I have learned a great deal from the process, and it has reinforced in me the feeling that I am a member of a team engaged in rural revitalization. Mr. Furuta told me that I should "just go to the locality," and so I would like to increase the number of times I visit the areas involved.

Sugata: It is by visiting the areas involved that you can experience the concrete aspects, and the unique atmosphere that touches all five senses. This is something we must bear in mind, as consultants who tend to be abstract.

Murakami: The last presenter, Mr. Maki, said that it is important to proceed as if you were building a large bonfire and gradually add wood to the embers. You could say that the past 18 months have been the warm-up stage for this Trailblazers Conference, as well as for the Intelligence Platform.

Nakata: That's right. This kind of perspective is usually not seen at consulting firms. They tend to get so caught up trying to produce results as quickly as possible in the short term that they lose sight of pursuing value in the long run. As a result, not much has been achieved yet in terms of rural revitalization. I feel that those of us at EY have been reminded of the need to return to our philosophy of long-term value.

Moving on to develop the next generation: the second stage of the Trailblazers Conference

——What should we do during the next stage? What are your plans for the future?

Nakata: EY must also consider ways to link the conference to business. This will be the stage where we will consider a structure to horizontally connect the human relationships that developed during warm-up operations and build a strategy as an organization.

Sugata: Not only rural revitalization, but social issues, too, cannot be solved just by the power of a few people or organizations. An issue-driven approach is required in which all players come together to address a single issue to realize a mutual-aid society that does not rely solely on public aid through public enterprises or independent financing through private ventures. I have not yet seen a market there that can make money, but I want to be ready to jump in and create such a market.

Murakami: Regional economies basically tend to be order-based. They wait for orders from large cities and major companies. And they rely on public support to make up the shortfall. In other words, their businesses work because the shortfall in their revenue is compensated for by local tax subsidies from the national government. However, the Showa paradigm, which was based on a model of following orders and distribution, is no longer sustainable for local communities. This why the concept of mutual aid is important, and "select and invest" need to replace "follow and distribute."

It's quite likely that people in rural areas who are starting to realize this came to the symposium and talked to us. There must be many more like them all over Japan. We should be able to create a new venue to connect such people, and bring the theories we have formulated to the new venue to inspire them.

Sugata: That would give rise to people who will be the next generation of trailblazers. I think this is a very important step toward the social implementation of the rural revitalization model.

Murakami: Yes, if we call the current trailblazers generation 0, it would be good to gradually expand the circle of people to generation 1 and then generation 2. In other words, this will be a loop of activities to change the culture in regions to one of "selection and investment."

At the same time, another major proposition remains. This is the problem that a business created through selection and investment itself will not be sustainable unless it is grown to a much larger scale. We have not yet found an effective solution to the question posed by Mr. Maki during discussions at the final session. As soon as a large amount of capital funding is brought in from the outside to expand the scale of the business, the local capital becomes the minority and the local side can no longer take ownership. But if the local side just carries on doing what it has been doing, the business itself will not expand to a stable scale.

From here, it is difficult to take the next step. We hope to continue to study this as an extension of the Trailblazers Conference.

Nakata: For this reason, the value of a region saying that they are keen to work together is really quite significant. We would like to reach a stage where we can enter a region from a stance of working together rather than an attitude of receiving work.

Sugata: Yes, as one team. In that case we will still be able to associate with the trailblazers. I would like to end by asking Mr. Hashimoto to say a few words of thanks, incorporating our hope for the further expansion of this "community of trailblazers" that resulted from Mr. Hashimoto's connections.

Hashimoto: Mr. Furuta, Mr. Takemoto, Mr. Kamiyama, Ms. Fujisawa, Mr. Kado, Mr. Abe, Mr. Maki, and Mr. Hori, Ms. Miyase, and Mr. Murakami, thank you very much. It was a wonderful conference. We will keep going as long as we hear the words "This isn't the end, is it?" I look forward to working with you again in the future.

Notes

Introduction

Special page on the *Ridilover Journal* website https://journal.ridilover.jp/projects/watakotopress

Session 1: Mr. Hima Furuta's Case—Mitoyo, Kagawa Prefecture

Chichibugahama Beach A beach in the Nio-cho district of Mitoyo, Kagawa Prefecture, that faces the Hiuchi-nada Sea. This is a popular spot for taking photos, especially for sharing on social media, because the surface of the water reflects the sky like a mirror, much like the "Salar de Uyuni" salt flats in Bolivia do.
Udon House A lodging facility converted from an abandoned *kominka* (old traditional Japanese house), where visitors can learn to make Sanuki udon noodles through hands-on workshops. They offer a Two-Day Program (day 1: Learn to make udon noodles and harvest vegetables from local farms; day 2: Visit two popular udon restaurants in the morning following an overnight stay.). https://udonhouse.jp/
Park-PFI The Park Private Finance Initiative (Park-PFI) is a public bidding system for selecting private-sector businesses to install facilities that will improve the appeal and convenience of urban public parks. Utilizing private-sector expertise, this initiative has developed facilities in numerous urban parks; these include cafes and shops that attract customers as well as nurseries and elder-care centers that cater to local residents.
Source: Guidelines for Utilizing Park-PFI Initiatives to Improve the Quality of Urban Public Parks (Ministry of Land, Infrastructure, Transport and Tourism (MLIT), City Bureau, Parks, Green Spaces, and Landscape Division)
Soichiro Coffee & Soychiro Tofu Soichiro Coffee is a take-out coffee shop, while Soychiro Tofu sells tofu and other groceries. Both of these businesses were founded by Mr. Soichiro Imagawa, who was born and raised in Mitoyo, with the goal of strengthening the local community.
Basic Infrastructure Model Under a "basic income" system, the government provides every citizen with the funds necessary to guarantee a minimum standard of living, regardless of their age, gender identity, or annual income. Conversely, a "basic infrastructure" system allows residents to enjoy the minimum standard of lifestyle services, such as food, clothing, and housing, as long as they contribute a certain amount of money. Mitoyo has proposed the latter system, and is currently working to establish it.
Source: City of Mitoyo homepage. https://basicmitoyo.jp/

Session 2: Mr. Yoshiteru Takemoto's Case—Ama, Shimane Prefecture

Dozen High School Miryokuka [revitalization] Project Oki-Dozen launched the first "High School Miryokuka Project" for making high schools more attractive, which is currently gaining popularity throughout Japan. Dozen High School was continually losing students, and the town would fall into further decline if it were to close. To address this impending danger, the local community launched a collaborative project with the prefectural high school, which 10 years later has set a precedent for the entire nation.
Source: High School Miryokuka Project homepage. http://miryokuka.dozen.ed.jp/
Dozen High School Dozen High School recruits students from across the country and around the world for their "island study" program. They've already hosted over 200 students from as far north as Hokkaido and as far south as Kagoshima. Some students enter Dozen High School after attending junior high abroad, while others transfer in from combined junior-senior high schools.
Source: Shimane Prefectural Oki-Dozen Senior High School homepage. https://www.dozen.ed.jp/
Half-Civil Servant, Half-X This refers to a unique way of working in Ama, in which one makes use of their interests and skills to give back to the local community as an "X" while also working for the town office as a civil servant.
Source: "*Nai Mono wa Nai*" Ama Town Official Note. https://ama-town.note.jp/
Future Co-Creation Fund Ama created this fund by donating the capital generated by the Hometown Tax Program to the Ama Future Investment Committee in order to support the future of the island by revitalizing local industries and creating increased synergy between job creation and workforce development. Investments are made only in businesses that shape the future of Ama. Island residents and collaborators have many opportunities to provide input on the management of the fund, which has a lower limit of 5 million yen.
Multiple Work Cooperative This cooperative was formed by businesses seeking to implement a cross-organizational "multiple work" style that would allow people to combine various jobs on the island with different peak seasons, changing workplaces depending on the season, in order to increase the number of people who can enjoy working in the island's unique industries. This working style has been labeled "*amu* work," meaning "knitting work," as people can combine different jobs and "knit" them together in their own ways.

Session 3: Mr. Yasuhiro Kamiyama's Case—Institutionalizing Private Lodging

DMO DMO stands for "Destination Management/Marketing Organization." These companies lead the way in managing and marketing tourist destinations, adopting a scientific approach that involves a wide variety of local stakeholders. The Japan Tourism Agency, an arm of the national government, has created an official registration system for DMOs that meet its requirements. These DMOs will provide guidance for developing and managing tourist destinations in ways that draw out their 'earning power' and foster both pride and affection. They also serve the purpose of coordinating the formulation and implementation of strategies for developing tourist destinations based on clearly defined concepts in cooperation with various stakeholders.

Entô Address: 1375-1 Fukui, Ama Town, Oki District, Shimane Prefecture
Renovated in 2021, Entô is a local hub where visitors can stay overnight in Ama, a town on Nakanoshima island in the Oki archipelago.
Source: Entô homepage. https://ento-oki.jp/

Private Lodging Business Act The Private Lodging Business Act was enacted in June 2017 to establish defined rules for the rapidly increasing "private lodging" (*minpaku*) facilities and promote the spread of upstanding private lodging services in response to the lack of safety and hygiene assurance, widespread neighborhood disputes over rising noise levels and improper garbage disposal, and the diversifying accommodation needs of tourists.

This law defines the roles and obligations of each of the three major players: private lodging business operators, private lodging administrators, and private lodging agents.

Special Zone Private Lodging Special exceptions to the Hotel Business Act, made in accordance with the National Strategic Special Zone Act, allow businesses to lodge foreign tourists in suitable private lodging facilities for longer than the period of time designated by their lease agreements and/or any supplementary agreements. This type of private lodging business has been operated in Ota Ward, Tokyo, as well as in Osaka City in Osaka Prefecture and other areas designated as National Strategic Special Zones. Unlike those operating under the Private Lodging Business Act, they are not limited to a certain number of business days per year.

National Strategic Special Zones The National Strategic Special Zone System was established to create "the world's most business-friendly environment" by implementing the bold regulatory and institutional reforms necessary to implement growth strategies. Comprehensive and intensive efforts were made to develop special exceptions to strict regulations that had not been revised for many years, and also to reform various other related systems.
Source: National Strategic Special Zones homepage. https://www.chisou.go.jp/tiiki/kokusentoc/index.html

Online Travel Agent (OTA) Websites where consumers can search for and purchase travel products and services offered by travel suppliers, such as lodging, airline tickets, car rentals, local tours, cruises, and other activities. Also called "Online Travel Agency."

***Albergo Diffuso* (AD): Decentralized hotels** An initiative originating in Italy that aims to combat depopulation caused by the declining birthrate and aging population by using tourism to solve the growing problem of vacant houses.

Vacant houses in participating communities are given new life as hotels, forming horizontal networks of integrated lodging facilities and restaurants, etc., centered around a core base that serves as the reception area. As such, the concept is sometimes referred to as a "community-wide hotel" in Japanese.
Ospitalità Diffusa **(OD): Dispersed hospitality** Translated literally, this term means "scattered hospitality." The basic idea is similar to that of AD, but while AD facilities are concentrated within a 200-meter radius of the reception area, OD facilities are dispersed over a wider area (generally within a one-kilometer radius) and integrated into a unified and continuous concept designed for sharing services and providing value to travelers.
Albergo Diffuso Town (ADT): Local Government Certifications Certifications granted by Albergo Diffuso Internazionale (ADI) to local governments that plan and promote AD and/or OD for the sustainability and development of their communities.
Source: Albergo Diffuso Internazionale Estremo Oriente (ADIeo). https://albergodiffuso.jp/
Northern Grande Hachimantai A new establishment that makes Hachimantai National Park feel just like a resort. With a dynamic view of the magnificent Mt. Iwate, Northern Grande Hachimantai offers a menu of locally produced ingredients for visitors to enjoy in a modern space that blends seamlessly with the surrounding nature.
Source: Northern Grande Hachimantai homepage. https://n-grande.com/atmosphere/
Hirado Castle Stay—*Kaiju Yagura* The *Kaiju Yagura* turret of Hirado Castle has been reborn as Japan's first permanent "castle stay" facility in Nagasaki, Kyushu. Guests can enjoy dinner and breakfast in the castle's spacious living and dining rooms, then look out over the ocean surrounding Hirado Island from the bathroom with three glass walls. Hirado Castle provides luxurious and novel encounters yet to be experienced by anyone else, creating the overwhelming impression of having escaped from ordinary day-to-day life, which serves as its key selling point.
Source: Hirado Castle Stay Kaiju Yagura. https://www.castlestay.jp/stay/

Session 4: Ms. Kumi Fujisawa's Case—Kamishihoro, Hokkaido

Kamishihoro Lifelong Learning Center Opened by the town of Kamishihoro to raise awareness among its residents of the importance of lifelong learning in order to help them revitalize the town as a whole. Several times a year, the Center gathers prominent leaders from around the country to offer public lectures on relevant topics to both local residents and those outside the community, including senior citizens. It is operated by Lifelong Learning Town Kamishihoro, a community development company established in 2017. http://kamishihoro-town.com/kamishihorojuku/

Kamishihoro Roadside Station https://karch.jp/michinoeki/
Karch Co., Ltd. Established in 2018 as a regional tourism trading company that accurately conveys the value of Kamishihoro, Karch Co., Ltd. works together with local residents to solve the issues affecting the area and turn Kamishihoro into a tourist destination that is both a nice place to live in and a nice place to visit. They are involved in various lines of business, including DMO, product development, lodging, and electricity retail.
Karch Co., Ltd. / Kamishihoro Tourism Business Creation DMO. https://karch.jp/
Kamishihoro Share Office An office that provides the multiple users sharing the same space with a new way of working, allowing them to share the countryside with the city. The Naitai Highland Farm and the mountains of Higashi-Taisetsu are visible from the window. https://www.kamishihoro.work/
Local Vitalization Cooperators The Local Vitalization Cooperator (LVC) program relocates people from urban areas to depopulated and otherwise disadvantaged areas in order to encourage newcomers to settle there permanently. Participants take part in cooperative activities that include supporting revitalization efforts by developing, selling and promoting local brands and products, working in the agriculture, forestry, and fishing industries, and supporting local residents. LVC members are appointed by the local government and generally serve terms lasting from one to three years.

Although the details of LVCs' activities, their terms of employment, and their benefits and compensation vary from one local government to another, the Ministry of Internal Affairs and Communications (MIC) provides financial assistance of up to 4.8 million yen per member to cover the expenses required to support their activities. Members can also take advantage of several support services during their tenures, including daily consultations with the support desk and the alumni network as well as various training programs designed for them. Further support is available for those starting or taking over a business after their term of service ends.

There were 6,447 active LCVs nationwide in 2022, and the MIC plans to further promote LCV initiatives in order to direct more newcomers to rural areas by achieving their stated goal of enrolling 10,000 active members by 2026.
Ministry of Internal Affairs and Communications | Community Strength Building & Rural Revitalization | Local Vitalization Cooperators. https://www.soumu.go.jp/main_sosiki/jichi_gyousei/c-gyousei/02gyosei08_03000066.html
Smart Logistics Initiative Involving Drones Press Release from Aeronext, Inc. | Pilot Program of Advanced Drone Delivery in Kamishihoro (Tourism Product Development/First Drone Home Delivery in Japan/First Drone Delivery of Cattle Testing Specimens in Japan). https://prtimes.jp/main/html/rd/p/000000063.000032193.html
Nippo House Kamishihoro Having been chosen as an "SDGs Future City" for its support of sustainable development goals (SDGs), Kamishihoro attracts many

people working in new fields, such as demonstration experiments and other advanced initiatives, as well as new brands created in partnership with local businesses. This facility caters to those people, serving as a vacation home, a gateway to the local community, and a place to disseminate information about Kamishihoro. Designed by MUJI House alongside Yataro Matsuura, this house in Kamishihoro is open to everyone. https://kamishihoro.today/about/#Info

Session 5: Mr. Shintaro Kado's Case—Matsuyama, Ehime Prefecture

Machi-Pay A regional currency that can be used for shopping, dining, and leisure activities throughout Matsuyama, Machi-Pay combines electronic payments with a shared point system and coupons for the local shopping district. Machi-Pay is available as a mobile application and as a card, both of which are convenient to use.
Machi-Pay Official Website | A Smart Payment Service for Everyone from Matsuyama City, Ehime Prefecture. https://machica.jp/
Antler Cohort Program The Antler Cohort Program is one of the largest start-up support programs in the world. It runs in 25 cities across the globe, with 80,000 potential entrepreneurs having applied to date. Participants commit themselves to the program full-time for about 10 weeks in order to launch their own business by the time it ends.
Antler's Start-Up Support Program, the Antler Cohort Program, Currently Recruiting 2nd Session Applicants. https://www.antler.tokyo/

Session 6: Mr. Toshiki Abe's Case—Echigo-Tsumari, Niigata Prefecture

Echigo-Tsumari Art Triennale One of the world's largest international art festivals, The Echigo-Tsumari Art Trienniale is a pioneer of the regional art festivals held throughout Japan. The concept—taking a journey through the *satoyama* (area between foothills and flat land, used for agriculture and forestry) with art as a guide—is attracting attention both domestically and abroad as a leading example of arts-based community development. https://www.echigo-tsumari.jp/about/
SaaS (Software as a Service) SaaS refers to software (mainly application software) and/or a delivery format that makes necessary functions available as a service for as long as necessary. Generally, these services are multi-tenant single systems that use the Internet for their necessary functions. https://ja.wikipedia.org/wiki/SaaS
Regional Revitalization Entrepreneurs Regional Revitalization Entrepreneurs are employees of companies located in the three major metropolitan areas—Tokyo, Osaka, and Nagoya—who spend a certain period of time working for local governments in rural areas, utilizing their know-how and expertise to enhance the area's unique charm and value, revitalize the local economy, and work to provide security and safety. The Ministry of Internal Affairs and Communications (MIC) provides the necessary support for this initiative so that local governments and

companies can work together to direct more people to rural areas.
Ministry of Internal Affairs and Communications | Community Strength Building & Rural Revitalization | Regional Revitalization Entrepreneurs. https://www.soumu.go.jp/main_sosiki/jichi_gyousei/c-gyousei/bunken_kaikaku/02gyosei08_03100070.html
Social Impact Bond (SIB) A social impact bond, or SIB, is a form of public–private partnership in which administrative services are outsourced to private-sector NPOs and companies that carry out projects using funds procured from private investors. The government only pays these investors when their projects produce the previously agreed-upon results. Implementing projects that reduce social costs with private-sector funds also reduces administrative costs, and the investors receive a return. However, the government does not pay investors anything if the agreed-upon goals are not met. Important factors include the nature of the social issues being targeted, the businesses implementing the measures, the goals being set, the evaluating organization, and the intermediary support organizations managing them all. SIBs, which combine results-based payments from businesses and the government with the use of private-sector funds, constitute one type of results-based outsourcing contract alongside results-based payments that do not use private funds.

Session 7: Mr. Daisuke Maki's Case—Nishiawakura, Okayama Prefecture

100-Year Forest Vision A village revitalization project based on the vision of the then-mayor that revolves around the forests of Nishiawakura, which chose to remain independent amidst the marked rise in municipal mergers during the Heisei era.
The 100-Year Forest Vision—Nishiawakura Town Hall. https://www.vill.nishiawakura.okayama.jp/
Nishiawakura Forest School (*Mori no Gakko*) The Nishiawakura Forest School will never stop believing in the potential of people and their communities. They work in manufacturing, with the aim of creating value by combining human labor with local resources. The Nishiawakura Forest School started in the staff room of a shuttered elementary school. Now, 12 years on, they process lumber, grow strawberries, operate their own café, and run nature experience programs.
Homepage | Nishiawakura Forest School. https://morinogakko.jp/
Nishiawakura Local Venture School A place for future entrepreneurs willing to walk forward on their own two feet, based in Nishiawakura, a village in Okayama Prefecture with a population of 1,400 people. Here they can take action based on their own business concept and vision, and strive to achieve more than just independence for their businesses by making use of training, conversations, and opportunities to connect with the Nishiawakura Village Office as well as other successful entrepreneurs. Support alone does not make entrepreneurs succeed; first, they have to act of their own accord to achieve the business goals that they've

set for themselves. The Nishiawakura Local Venture School aids them in these endeavors by giving them opportunities to improve their precision and broaden their horizons.
The Nishiawakura Local Venture School is Now Accepting Applications for 2021! [Want to start a rural revitalization business in Okayama?]. https://www.throughme.jp/lvs-lp/venture.html
Takibi ("bonfire") Program An entrepreneurship program in which participants workshop business topics that appear somewhat promising for the community, sharpen their skills by working with interns and pro bono volunteers (people who contribute by freely offering their knowledge and skills), and launch their companies with players prepared to move forward in earnest.
Special Feature: "SDGs x Enjoying Life 2021" 4 Takibi Program—Launch of the Takibi Program for Building New Economies in Local Communities! | Through Me. https://throughme.jp/idomu_nishiawakura_sdgs_04/
Forest GIS The Forest Geographic Information System (GIS) digitally processes basic data about the forest, such as basic forest maps, forest planning maps, and forest registers, thereby centralizing the management of maps and registers that were previously managed individually. This system can also link basic forest data from public organizations (prefectural and local governments) with location data utilizing GIS maps, which enables users to quickly understand the current status of the forest and promptly respond by performing a variety of related tasks. https://fgis.jp/
Through Me Through Me is a regional media company run by A-Zero Group, Inc., which was founded by Mr. Daisuke Maki. They aim to fulfill their goal of inspiring one million people to live true to themselves by learning about the lifestyles of people from remote areas all around the country, gathering their wisdom, and reporting on the diverse sets of values that they hold. Through Me documents the efforts of people in the village of Nishiawakura and the town of Atsuma, among other places, to revitalize their local communities. https://throughme.jp/

Profiles

Toshiki Abe

Born in 1987, Abe ran away from home at age 14 after a violent outburst. He became a juvenile delinquent and stopped attending school, but with the support of his friends, he made up his mind to enter Tokyo University. While enrolled there, he founded Ridilover as a platform for study tours that allowed participants to experience the realities of current social issues firsthand. Abe has won numerous titles and awards, including the Student Entrepreneur Championship and the Grand Prize in the 5th KDDI Mugen Labo Awards, and was selected for the Forbes Asia 30 Under 30 List. He is the author of the book *To You, Our Future Leaders* (Nikkei BP, 2015), among others.

Kumi Fujisawa

After graduating from Osaka City University, Fujisawa worked for several domestic and international investment fund management companies before founding Japan's first investment trust evaluation company in 1995. In 1999, she sold the company to S&P Global Ratings, the world's leading rating agency. The following year, she co-founded Think Tank SophiaBank, which she led from 2013 until March of 2022. In 2007, Fujisawa was chosen by the World Economic Forum, which presides over the Davos Conference, as one of their Young Global Leaders. In 2008, she was appointed to their Global Agenda Council, which discusses global issues, and visited over 40 countries during her tenure as a member. Fujisawa serves on the advisory boards of several government ministries, as well as those of public service corporations such as the Japan Securities Dealers Association and the Japan Professional Football League. She is also an outside director for several publicly listed companies, including Shizuoka Bank Ltd. and the Toyota Tsusho Corporation. Making use of her own entrepreneurial experience, Fujisawa has reported on small and medium-sized businesses nationwide as an anchor for the "21st Century Business School" program on NHK Educational TV. She has actively created opportunities to meet and speak with numerous leaders, both in Japan and abroad, and has covered over 1,000 companies. In March 2020, she graduated from the Waseda University Graduate School of Sport Sciences at the top of her class. In April 2022, she was appointed chairperson of the Institute for International Socio-Economic Studies, Ltd, a non-partisan think tank founded by the NEC Group.

Hima Furuta

Furuta was born in Tokyo. He attended Keio University, but withdrew before graduation. Furuta has been involved in numerous community production and corporate branding projects, including the Marunouchi Morning University in Marunouchi, Tokyo. He has worked to bridge the gap between cities and rural areas, as well as between Japan and other countries, through projects that include the agricultural experiment restaurant Roppongi Nouen, the Peace Kitchen Project that aims to spread Japanese cuisine throughout the world, and the Udon House, an inn that promotes the culture of Sanuki udon noodles. Currently, Furuta is investing in startups that strive to effect change in their local communities and in society at large. He serves as an advisor to highway bus company Willer Express, which also operates restaurant buses; the crowdfunding service Campfire; and the renewable energy company Shizen Energy Inc. He has also been appointed to the board of a medical corporation.

Shintaro Kado

Kado was born in Minatomachi, Matsuyama, in 1982. After graduating from the Keio University Faculty of Economics, he was hired by the investment banking firm Goldman Sachs. Later, he returned to his hometown and took over the family apparel business. In 2014, Kado was appointed chairperson of the Matsuyama Gintengai Shopping Center Promotion Association. That same year, he became president and representative director of Machidukuri-Matsuyama Co. Ltd., while also serving as vice-chairperson of the Ehime Prefecture Federation of Shopping Center Promotion Associations and chairperson of the Ehime Prefecture Youth Federation of Shopping Center Promotion Associations. In 2015, Kado was made chairperson of the Institute Ojoka-Matsuyama, and in 2016, he joined Ehime FC Co., Ltd. as a board member. In 2017, he was appointed head of the Youth Division of the National Federation of Shopping Center Promotion Associations, and then joined the board of Ehime Sports Entertainment Co., Ltd. in 2019. Since September of 2021, he has served as vice-chairperson of the National Federation of Shopping Center Promotion Associations and chairperson of the Ehime Prefecture Federation of Shopping Center Promotion Associations.

Yasuhiro Kamiyama

Kamiyama was appointed executive officer in charge of new business at Rakuten Travel, Inc. in September 2007, after serving as director and general manager of the business division at KLab, Inc. He left Rakuten Travel in 2012 and founded Hyakusenrenma, Inc. that June, where he serves as president and representative director. He has also worked as a part-time lecturer at the Tokyo Metropolitan University, served on the Expert Council for the Utilization of Historical Resources under the Cabinet Secretariat of Japan, and been appointed board chairperson of the Japan Countryside Stay Association.

Daisuke Maki

Maki was born in Kyoto, and graduated from the Kyoto University Graduate School of Agriculture with a major in Forest Ecology. After working at a private think tank, he became the director of the AMITA Institute for Sustainable Economies, which he co-founded in 2005. He has launched many new business projects in agricultural, mountain, and fishing villages, including the improvement of forestry management using the FSC certification system. In 2009, Maki founded the Nishiawakura Forest School to process and distribute lumber. In 2015, he founded A-Zero Group, Inc., and began research and development into the comprehensive diversification of the agriculture, forestry, and fishing industries.

Keisuke Murakami

Murakami serves as the director-general of the Digital Agency Public Service Group. He was born in Tokyo in 1967, and began working for the former Ministry of International Trade and Industry (MITI) in 1990. He has been involved in shaping IT policies, launching the Cool Japan strategy, conducting international negotiations on global warming at COP15 and 16, and establishing a feed-in tariff system for renewable energy. Since 2014, Murakami has been working on rural revitalization strategies under the Cabinet Secretariat and Cabinet Office. In July 2020, he was appointed head of the Management Support Department of the Small and Medium Enterprise Agency. He was appointed as councilor of the National Strategy Office of Information and Communications Technology under the Cabinet Secretariat in July 2021, and then to his current position that September.

Yoshiteru Takemoto

Takemoto was born in 1971. After completing graduate school, he worked at a foreign accounting firm and a foreign think tank, and in 2009 he founded Tobimushi, Inc. with the aim of establishing a system that would circulate materials, energy and food within the region while revitalizing local forestry. Takemoto's specialty is environmental law, and he has supported draft multiple environmental policies in Japan. Additionally, he has extensive experience with developing multifaceted environmental businesses, owing to his knowledge of finance and financial accounting, and is familiar with the actual scene of both legislation and entrepreneurship.

Jun Hori

Hori is the representative director of the Watashi o Kotoba ni Suru Research Institute. He was born in Kobe City in 1977. He studied media while at university and worked for the NHK after graduation. Currently, he works as a freelance journalist. He specializes in using both words and images to communicate, and has also produced documentary films. He has traveled around the world, including to Africa, the Middle East, Asia, Europe, and the United States, to interview the people living there about their ideas of happiness. He believes that being specific is more important than making sweeping generalizations.

Mayuko Miyase

Miyase is the director of the Watashi o Kotoba ni Suru Research Institute. She was born in Fukuoka Prefecture in 1982. Having now lived in Tokyo for over half of her life, she wants to rediscover the charms of Fukuoka, her hometown. Miyase previously worked for Fuji Television, and is currently a freelance announcer. She wants to use her words to share wonderful things and empathize with others, and she values the realizations that come from her hobby of observation.

EY Strategy and Consulting Co., Ltd.
A firm with a consulting service line that handles management consulting for EY Japan and a strategy and transactions service line that supports strategic transactions. EY's diverse team of experts collaborate with over 360,000 members in more than 150 countries and regions around the world to provide one-stop support for strategy-to-execution (M&A) and strategy-to-transformation.

Trailblazers Conference—EY Intelligence Platform
Repository of videos and minutes from every meeting. https://go.ey.com/35iUQKI
Mitsuhiro Sugata, Partner, Japan Region; Non-audit Services Markets Leader
Hiroyuki Nakata, Partner, Government and Infrastructure Unit
Hiroko Hashimoto, Manager, Government and Infrastructure Unit

The logomark of EY Intelligence Platform

Left to right: Hiroyuki Nakata, Mitsuhiro Sugata, and Hiroko Hashimoto.

Rural Revitalization Trailblazers Conference

Conference Members (Trailblazers):
Toshiki Abe (representative director, Ridilover Inc.)
Kumi Fujisawa (chairperson, Institute for International Socio-Economic Studies, Ltd.)
Hima Furuta (representative director, Umari Inc.)
Shintaro Kado (president and representative director, Machidukuri-Matsuyama Co., Ltd.)
Yasuhiro Kamiyama (president and representative director, Hyakusenrenma, Inc.)
Daisuke Maki (president and representative director, A Zero Group inc.)
Keisuke Murakami (director-general of the Digital Agency, Government of Japan)
Yoshiteru Takemoto (representative director, tobimushi Inc.)

Facilitators:
Jun Hori (representative director, Watashi o Kotoba ni Suru Research Institute)
Mayuko Miyase (director, Watashi o Kotoba ni Suru Research Institute)

Writer:
Ichiro Matsuoka (e-script Corporation)